Recent Titles in This Series

49 **Robert L. Devaney, editor,** Complex dynamical systems: The mathematics behind the Mandlebrot and Julia sets (Cincinnati, Ohio, January 1994)

48 **Walter Gautschi, editor,** Mathematics of Computation 1943–1993: A half century of computational mathematics (Vancouver, British Columbia, August 1993)

47 **Ingrid Daubechies, editor,** Different perspectives on wavelets (San Antonio, Texas, January 1993)

46 **Stefan A. Burr, editor,** The unreasonable effectiveness of number theory (Orono, Maine, August 1991)

45 **De Witt L. Sumners, editor,** New scientific applications of geometry and topology (Baltimore, Maryland, January 1992)

44 **Béla Bollobás, editor,** Probabilistic combinatorics and its applications (San Francisco, California, January 1991)

43 **Richard K. Guy, editor,** Combinatorial games (Columbus, Ohio, August 1990)

42 **C. Pomerance, editor,** Cryptology and computational number theory (Boulder, Colorado, August 1989)

41 **R. W. Brockett, editor,** Robotics (Louisville, Kentucky, January 1990)

40 **Charles R. Johnson, editor,** Matrix theory and applications (Phoenix, Arizona, January 1989)

39 **Robert L. Devaney and Linda Keen, editors,** Chaos and fractals: The mathematics behind the computer graphics (Providence, Rhode Island, August 1988)

38 **Juris Hartmanis, editor,** Computational complexity theory (Atlanta, Georgia, January 1988)

37 **Henry J. Landau, editor,** Moments in mathematics (San Antonio, Texas, January 1987)

36 **Carl de Boor, editor,** Approximation theory (New Orleans, Louisiana, January 1986)

35 **Harry H. Panjer, editor,** Actuarial mathematics (Laramie, Wyoming, August 1985)

34 **Michael Anshel and William Gewirtz, editors,** Mathematics of information processing (Louisville, Kentucky, January 1984)

33 **H. Peyton Young, editor,** Fair allocation (Anaheim, California, January 1985)

32 **R. W. McKelvey, editor,** Environmental and natural resource mathematics (Eugene, Oregon, August 1984)

31 **B. Gopinath, editor,** Computer communications (Denver, Colorado, January 1983)

30 **Simon A. Levin, editor,** Population biology (Albany, New York, August 1983)

29 **R. A. DeMillo, G. I. Davida, D. P. Dobkin, M. A. Harrison, and R. J. Lipton,** Applied cryptology, cryptographic protocols, and computer security models (San Francisco, California, January 1981)

28 **R. Gnanadesikan, editor,** Statistical data analysis (Toronto, Ontario, August 1982)

27 **L. A. Shepp, editor,** Computed tomography (Cincinnati, Ohio, January 1982)

26 **S. A. Burr, editor,** The mathematics of networks (Pittsburgh, Pennsylvania, August 1981)

25 **S. I. Gass, editor,** Operations research: mathematics and models (Duluth, Minnesota, August 1979)

24 **W. F. Lucas, editor,** Game theory and its applications (Biloxi, Mississippi, January 1979)

23 **R. V. Hogg, editor,** Modern statistics: Methods and applications (San Antonio, Texas, January 1980)

22 **G. H. Golub and J. Oliger, editors,** Numerical analysis (Atlanta, Georgia, January 1978)

21 **P. D. Lax, editor,** Mathematical aspects of production and distribution of energy (San Antonio, Texas, January 1976)

20 **J. P. LaSalle, editor,** The influence of computing on mathematical research and education (University of Montana, August 1973)

(Continued in the back of this publication)

AMS SHORT COURSE LECTURE NOTES
Introductory Survey Lectures
published as a subseries of
Proceedings of Symposia in Applied Mathematics

Proceedings of Symposia in
APPLIED MATHEMATICS

Volume 47

Different Perspectives on Wavelets

American Mathematical Society
Short Course
January 11–12, 1993
San Antonio, Texas

Ingrid Daubechies
Editor

American Mathematical Society
Providence, Rhode Island

LECTURE NOTES PREPARED FOR THE
AMERICAN MATHEMATICAL SOCIETY SHORT COURSE

WAVELETS AND APPLICATIONS

HELD IN SAN ANTONIO, TEXAS
JANUARY 11–12, 1993

The AMS Short Course Series is sponsored by the Society's Program Committee for National Meetings. The series is under the direction of the Short Course Subcommittee of the Program Committee for National Meetings.

1991 *Mathematics Subject Classification.*
Primary 35A27, 42C15, 46E15, 62A99, 94A11.

Library of Congress Cataloging-in-Publication Data

Different perspectives on wavelets/Ingrid Daubechies, editor.
    p.   cm. — (Proceedings of symposia in applied mathematics, ISSN 0160-7634; v. 47)
    Includes bibliographical references.
    ISBN 0-8218-5503-4 (acid free)
    1. Wavelets.   I. Daubechies, Ingrid.   II. Series.
QA403.3.D54   1993
531′.1133–dc20

93-33264
CIP

All articles in this volume were printed from copy prepared by the authors.
The articles were typeset using AMS-LATEX,
the American Mathematical Society's TEX macro system.

10 9 8 7 6 5 4 3 2   00 99 98 97 96 95

Table of Contents

Preface   ix

Wavelet Transforms and Orthonormal Wavelet Bases
INGRID DAUBECHIES   1

Wavelets and Operators
YVES MEYER   35

Projection Operators in Multiresolution Analysis
PIERRE GILLES LEMARIÉ-RIEUSSET   59

Wavelets and Differential Operators
PHILIPPE TCHAMITCHIAN   77

Wavelets and Fast Numerical Algorithms
GREGORY BEYLKIN   89

Wavelets and Adapted Waveform Analysis. A Toolkit for Signal Processing and Numerical Analysis
RONALD R. COIFMAN AND M. VICTOR WICKERHAUSER   119

Best-adapted Wavelet Packet Bases
MLADEN VICTOR WICKERHAUSER   155

Nonlinear Wavelet Methods for Recovery of Signals, Densities, and Spectra from Indirect and Noisy Data
DAVID L. DONOHO   173

Preface

With hindsight the wavelet transform can be viewed as a synthesis of ideas that have emerged since the 60-s (and for some aspects even earlier) in fields as diverse as mathematics (pure as well as applied), physics and electrical engineering. The basic idea is always to use a family of building blocks to represent the object at hand (a function, an operator, a signal or image,...) in an efficient and/or insightful way; the building blocks themselves come in different sizes, and are suitable for describing features with a resolution commensurate with their size. This no doubt sounds rather vague, and the different contributions in this book flesh it out in different ways.

The first chapter in this book is the most introductory. It gives some motivation for the wavelet transform (although the proof of the pudding is in the eating – the true motivation is not given by abstract considerations, but by the usefulness of wavelets in proving real-life theorems or analyzing real-life data; some such applications are the main message of later chapters) and it describes various different types of wavelet transform, with special emphasis on orthonormal wavelet bases, because that is what most of the other chapters need.

There are two important aspects to wavelets, which I shall call "mathematical" and "algorithmical". In their mathematical aspect, wavelets are rooted in the use of dilations and convolutions in Calderón-Zygmund theory in harmonic analysis. The techniques used and refined there in the past 25 years had led to powerful tools suited to proving hard theorems. But that was the full extent of their range; they had not led to any applications in numerical analysis or signal analysis before the advent of wavelets. Algorithmically, wavelets are related to subband filtering in electrical engineering. Subband filtering was developed from the 70-s on; exact reconstruction procedures were discovered in the early 80-s. These were obviously fast algorithms, meant as a front-end processing step before encoding or compressing information in various types of signals. A lot of effort went into optimizing the filters for various applications, and this subfield of electrical engineering is now quite mature. But the results were purely algorithmic; they were never intended or viewed as a powerful mathematical tool that could interest people other than signal processors. Another algorithmic ancestor of wavelets are the multiple algorithms in numerical analysis, closer to mathematics, but still ad hoc.

Wavelets then incorporate both aspects: the fast algorithms of subband coding and the powerful mathematical potential of the Calderón-Zygmund theory tools. Chapters 2 to 8 in this book all illustrate different aspects of (bases of) wavelets, some more mathematical, others more algorithmical. Chapters 2 to 4 are mostly from the mathematical point of view (although the ideas in chapters 3 and 4 are being implemented in numerical work, which would be impossible without the fast algorithms); chapters 5 to 7 place more emphasis on the algorithmic aspect, without losing sight of the mathematical properties which are essential for any of these applications to even make sense. Chapter 8 combines some of the deep mathematical properties of wavelets (their "adaptability" to many functional spaces) with their algorithmic ease for still different applications. All the chapters in this volume have been written for these short course notes and are in final form as such; it should be noted, however, that each constitutes the summary of several mathematical research articles, some of which still have to appear.

Even though the first chapter explains many of the basic properties of wavelets in detail, most chapters contain their own short review of wavelet bases. I have not tried to edit away this slight overlap: the small differences in emphasis on which properties are deemed most important by the different authors is an illustration of the versatility of wavelets. All the authors chose also the notation that suited their purpose best; the notation is therefore not uniform throughout the book, but every chapter defines its own notation carefully.

I am very glad that some of the chapters (especially chapter 6) contain constructions "beyond wavelets" (wavelet packets, local trigonometric bases). Wavelets are an exciting development, and they let us put a foot in many doors, but they are not the answer to everything. And even for problems where they seem perfect, we need to develop refined and efficient tools to make use of their properties, such as is done in chapters 4 and 5. This book is therefore a good illustration of the powerful potential of wavelets in different directions, a potential that we hope some of our readers will help develop!

I should add that there are many very interesting developments using wavelets that were not represented at the short course and are not included in these lecture notes – including some of my favorites. But there was only so much we could do in one short course, or in one volume of course notes.

I would like to use this opportunity to thank all the speakers for their participation. It was not easy to make a short course where everybody would find something to like, from wavelet novices to experts working with them for the past four years, and I think we succeeded. (Although the result was also a course where everybody found something to dislike!) I would like to thank especially Yves Meyer for still giving us his contribution for this volume, even though he had to cancel his lecture, and Pierre-Gilles Lemarié-Rieusset for substituting for Yves Meyer at very short notice.

Finally, I would like to thank Tina Sharp for the seemingly impossible task of massaging all the faxes, typescripts, diskettes, e-mail files and scribbled notes into one coherent whole.

Ingrid Daubechies
AT&T Bell Laboratories
Murray Hill, NJ 07974
June 30, 1993

Proceedings of Symposia in Applied Mathematics
Volume **47**, 1993

Wavelet Transforms and Orthonormal Wavelet Bases

INGRID DAUBECHIES

ABSTRACT. We introduce the wavelet transform and discuss its motivation as a time-frequency localization tool. We review the different types of wavelet transform, with a special emphasis on orthonormal wavelet bases and their properties. We finish by a short discussion of their shortcomings.

"Wavelets" or "wavelet transforms" are a tool for decomposing functions in various applications, several of which are presented in this short course. The functions to be analyzed can be solutions of a differential equation with shocks, or integral kernels of singular integral operators, or 1 or 2-dimensional signals, as in sound (speech or music), time series or images. The wavelet transform can be viewed as a synthesis over the last fifteen years of ideas from many different fields, ranging from pure mathematics to quantum physics and electrical engineering. I will give here a description of several types of wavelet transform, with a special emphasis on (orthonormal) wavelet bases.

1. Time-frequency localization: what and why?

Let $f(t)$ be a function depending on time. If we are interested in its "frequency content" or "spectrum", our first reflex is to compute its Fourier transform,

$$\hat{f}(\xi) = \frac{1}{\sqrt{2\pi}} \int f(x) \, e^{-i\xi x} \, dx \, .$$

Just as the different harmonic components were present in $f(t)$, but impossible to read off at a glance, so the time information is present in $\hat{f}(\xi)$ but hard to

1991 *Mathematics Subject Classification*. Primary 33E20; Secondary 46E15, 41A15, 94A12.

These notes were written while the author was in the Mathematics Department at Rutgers University, partially supported by NSF grant 4-20875.

read off (it is all hidden in the phase of $\hat{f}(\xi)$). Often we would like to have a frequency decomposition of f locally in time, similar to music notation, which tells the musician which note (= frequency information) to play when (= time information). This is what is achieved by so-called time frequency representations. The wavelet transform of f can be viewed as such a time frequency representation. There exist other, older and very useful time frequency representations. The most widely used is the windowed Fourier transform. Here the function f is first "windowed" by multiplying it by a fixed $g(t)$ (the window function); this effectively restricts f to an interval (with smoothed edges) (see Figure 1). Then

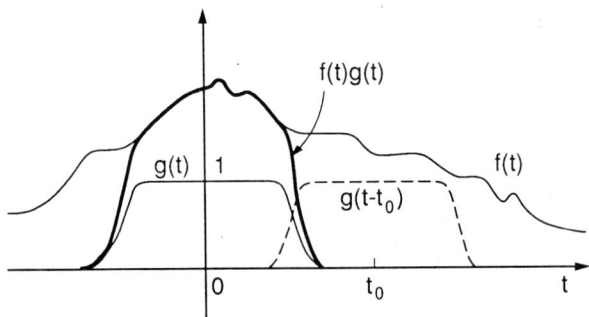

FIGURE 1.

the Fourier coefficients of this product are computed. This process is repeated with shifted versions of g, i.e. $g(t - nt_0)$, $n \in \mathbb{Z}$, leading to a family of windowed Fourier coefficients,

$$(1.1) \qquad S_{m,n}(f) = \int f(s)\, g(s - nt_0)\, e^{im\omega_0 s}\, ds\ ,$$

with $m, n \in \mathbb{Z}$. These can also be viewed as the inner products (in $L^2(\mathbb{R})$) of f with the

$$(1.2) \qquad g_{mn}(t) = e^{-im\omega_0 s} g(t - nt_0)$$

(we assume g is real). Each g_{mn} consists of an envelope function, shifted by nt_0, and then "filled in" with oscillations (see Figure 2); the index n gives us the time localization of g_{mn}, the index m its frequency.

The wavelet transform is similar to the windowed Fourier transform in that it also computes inner products of f with a sequence of functions $\psi_{m,n}$, with m indicating frequency localization, and n time localization,

$$(1.3) \qquad W_{m,n}(f) = \int f(s)\, \overline{\psi_{m,n}(t)}\, dt\ ,$$

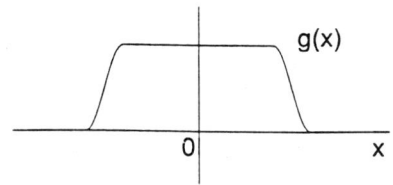
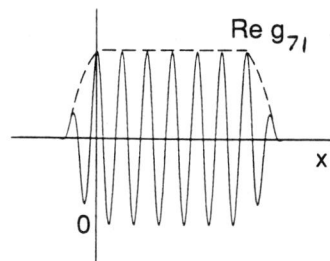

FIGURE 2.

but the $\psi_{m,n}$ are generated in different way. The basic wavelet ψ is typically well concentrated in time and in frequency, and has integral zero

$$(1.4) \qquad \int \psi(t)\, dt = 0 \; ,$$

which means it has at least some oscillations. The $\psi_{m,n}$ are then generated by dilations and translations:

$$(1.5) \qquad \psi_{m,n}(t) = a_0^{-m/2} \psi(a_0^{-m} t - n b_0) \; ,$$

where $a_0 > 1$ and $b_0 > 0$ are fixed parameters (similar to the ω_0, t_0 in (1.1), and m, n range over all of $\mathbb{Z}$). Changing m in (1.5) amounts to packing the oscillations of ψ into a smaller $(m > 0)$ or larger $(m < 0)$ width, i.e. to wavelets with higher or lower frequency ranges; for fixed m, the $\psi_{m,n}$ are then translates of $\psi_{m,0}$ by $n a_0^m b_0$, i.e. the wavelets are translated by amounts proportional to their width. A few typical wavelets are illustrated in Fig. 3. It is clear that high frequency

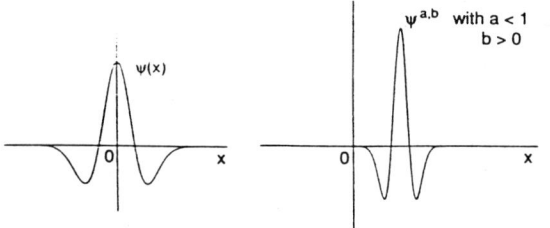
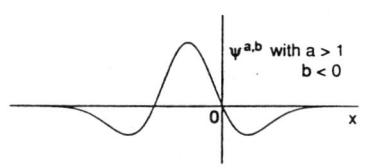

FIGURE 3.

wavelets are narrow, low frequency wavelets wide. This is the main difference between the wavelet transform and the short term Fourier transform: the $g_{m,n}$ of the short time Fourier transform all have the same width. It is therefore to be expected (and borne out in reality) that the wavelet transform is particularly well adapted to functions, signals or operators with highly concentrated high frequency components superposed on longer lived low frequency components.

2. Different types of wavelet transform.

2.1. The continuous wavelet transform. Here the dilation and translation parameters a, b vary continuously over $\mathbb{R}$. That is, we define (in 1 dimension; higher dimensional versions are straightforward)

$$(2.1) \qquad \psi^{a,b}(x) = a^{-1/2} \psi\left(\frac{x-b}{a}\right) ,$$

with $a, b \in \mathbb{R}$, $a > 0$. Then

$$(2.2) \qquad (Wf)(a,b) = \langle f, \psi^{a,b}\rangle = \int f(x)\, a^{-1/2}\, \overline{\psi\left(\frac{x-b}{a}\right)}$$
$$= \int \hat{f}(\xi)\, a^{1/2}\, \overline{\hat{\psi}(a\xi)}\, e^{ib\xi} ,$$

and

$$\int_0^\infty \int_{-\infty}^\infty \langle f, \psi^{a,b}\rangle \langle \psi^{a,b}, g\rangle\, db\, \frac{da}{a^2}$$
$$= 2\pi \int_0^\infty \int_{-\infty}^\infty \hat{f}(\xi)\, \overline{\hat{g}(\xi)}\, |\hat{\psi}(a\xi)|^2\, d\xi\, \frac{da}{a}$$
$$(2.3) \qquad = 2\pi C_\psi\, \langle f, g\rangle ,$$

provided that

$$(2.4) \qquad \int_0^\infty \xi^{-1} |\hat{\psi}(\xi)|^2\, d\xi = \int_{-\infty}^0 |\xi|^{-1}\, |\hat{\psi}(\xi)|^2\, d\xi =: C_\psi < \infty .$$

Condition (2.4) implies that $\int_{-\infty}^\infty |\xi|^{-1} |\hat{\psi}(\xi)|^2\, d\xi < \infty$, which (for reasonable ψ) amounts to the same as our earlier requirement $\int \psi(x)\, dx = 0$. Another ingredient in (2.4) is a symmetry of concentration in $|\hat{\psi}(\xi)|^2$, with respect to the measure $|\xi|^{-1}\, d\xi$, on positive and negative frequency axes. This requirement is automatically satisfied if ψ is real. On the other hand, if (2.3) is only required for real f, g, where, since $\hat{f}(-\xi) = [\hat{f}(\xi)]^*$, the positive frequency behavior completely determines the negative frequency analog, then one can find formulations in which the symmetry for $\xi \leftrightarrow -\xi$ in (2.4) is no longer necessary. Similarly, if one allows negative a in (2.1) and (2.2) then (2.4) collapses to $C_\psi := \int_{-\infty}^\infty |\xi|^{-1} |\hat{\psi}(\xi)|^2 < \infty$ (see Daubechies (1992)).

Formula (2.3) can also be rewritten as

$$(2.5) \qquad f(x) = \frac{1}{2\pi C_\psi} \int_{-\infty}^\infty \int_0^\infty \langle f, \psi^{a,b}\rangle \psi^{a,b}(x)\, \frac{da\, db}{a^2} ,$$

with weak convergence in L^2-sense. In fact, for reasonable ψ, (2.5) converges in many more topologies; in particular, it converges pointwise in any point x where f is continuous (see Holschneider and Tchamitchian (1990)).

Note that (2.5) can be read in two different ways: it tells us, once we know the $\langle f, \psi^{a,b} \rangle$, how to reconstruct f from these wavelet coefficients; it also gives a recipe for writing any arbitrary f as a superposition of $\psi^{a,b}$.

Formula (2.5) has in fact been known for quite a while: it is already implicit in Calderón (1964) as a useful mathematical tool (with completely different notations), and it appeared as the "reproducing identity for the $ax + b$-group" in Aslaksen and Klauder (1968). A similar and even older reproducing identity exists for the continuous windowed Fourier transform. (For an extensive discussion of these and other reproducing identities, see Klauder and Skagerstam (1985).)

It may seem puzzling that, according to (2.5), we can write any f, even if $\int f(x)dx > 0$, as a superposition of $\psi^{a,b}$, each of which has zero integral. The solution to this paradox is that (2.5) converges in L^2, or pointwise, but not in L^1. In fact, for any finite a_1, R, and any nonzero a_0, the functions

$$f_{a_0,a_1;R}(x) = \frac{1}{2\pi C_\psi} \int_{-R}^{R} \int_{a_0}^{a_1} \langle f, \psi^{a,b} \rangle \psi^{a,b}(x) \, \frac{da \, db}{a^2}$$

will have zero integral; for a_0 close to 0 and a_1, R very large, their graph will be very close to that of f, except that they will have large, shallow, negative "pools" in regions where f is small, leading to small pointwise or L^2 differences, but sufficient to ensure $\int f_{a_0,a_1;R}(x) \, dx = 0$. (See Figure 4.)

The continuous wavelet transform is useful when one wants to recognize or extract features. Scaling or translating f leads to a shift of the $(Wf)(a,b)$ in a and b, so that the whole analysis can be made to be scale and translation invariant, a desirable property in some applications. Of course, it can be cumbersome to have to deal with the very redundant $(Wf)(a,b)$: after all, we have changed a 1-dimensional function f into the 2-dimensional Wf; pictures of Wf may give insight into the different components of f, but this is only a first stage. Several groups, mostly in Marseille (France) have developed mathematical tools for extracting the "bare bones" from $Wf(a,b)$ and use these to describe f; an extensive review article is Delprat et al (1992).

2.2. The discrete but redundant wavelet transform: frames. The wavelet family (1.5) and the wavelet transform (1.3) can be viewed as discretized versions of the continuous wavelet transform, with a, b restricted to $a = a_0^m$; $b = nb_0a_0^m$.

In the discrete case, there does not exist, in general, a "resolution of the identity" formula analogous to (2.5) for the continuous case. Reconstruction of f from the $W_{m,n}(f)$, if at all possible, must therefore be done by some other means. The following questions naturally arise:

    (1) Is it possible to characterize f completely by knowing the $W_{m,n}(f)$?

    (2) Is it possible to reconstruct f in a numerically stable way from the $W_{m,n}(f)$?

These questions concern the recovery of f from its wavelet transform. We can also consider the dual problem, the possibility of expanding f into wavelets,

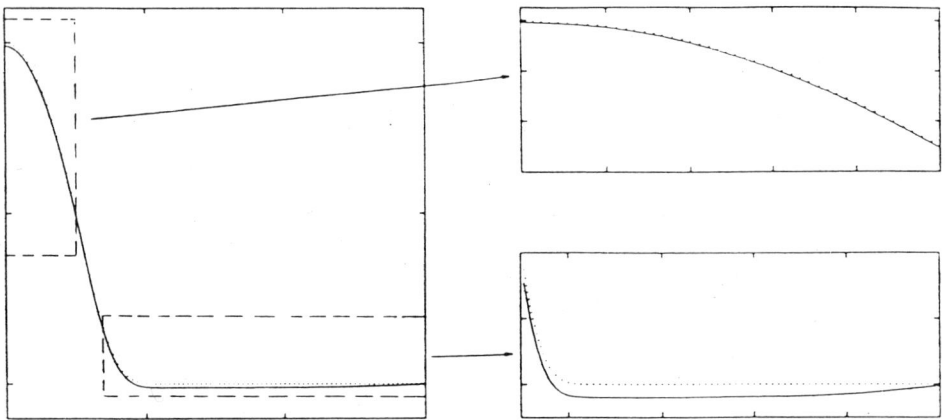

FIGURE 4. A Gaussian (dotted line) and its reconstruction (solid line) with cutoffs in a and b (see text). Only the right half is plotted; to make the effect more visible, two blowups are provided as well. The reconstruction has wide shallow "pools" on the side so that its integral is zero, even though it is close to the Gaussian at every point. Note that the horizontal scale is sinh x rather than x, giving a linear scale near 0 but an exponential scale further on.

which then leads to the dual questions:

(1′) Can any function be written as a superposition of $\psi_{m,n}$?

(2′) Is there a numerically stable algorithm to compute the coefficients for such an expansion?

As in the continuous case, these discrete wavelet transforms often provide a very redundant description of the original function. This redundancy can be exploited (it is, for instance, possible to compute the wavelet transform only approximately, while still obtaining reconstruction of f with good precision), or eliminated to reduce the transform to its bare essentials (such as in the image compression work of S. Mallat and S. Zhong (1992)). It is in this discrete form that the wavelet transform is closest to the "ϕ-transform" of Frazier and Jawerth (1988).

The choice of the wavelet ψ used in the continuous wavelet transform or in frames of discretely labelled families of wavelets is essentially only restricted by the requirement that C_ψ, as defined by (2.4), is finite. For practical reasons, one usually chooses ψ so that it is well concentrated in both the time and the frequency domain. For any such ψ, one can then find threshold values such that if a_0, b_0 are chosen below these thresholds, then all the questions above can be answered by "yes", and one can construct explicit algorithms. (For a much more extensive discussion, see Daubechies (1992).) All this still leaves a lot of freedom. Giving up a lot of this freedom allows one to build (orthonormal) bases of wavelets.

2.3. Orthonormal wavelet bases: the Haar basis as an example. For some very special choices of ψ and a_0, b_0, the $\psi_{m,n}$ constitute an orthonormal basis for $L^2(\mathbb{R})$. In particular, if we choose $a_0 = 2$, $b_0 = 1$, then there exist ψ, with good time-frequency localization properties, such that the

$$(2.6) \qquad \psi_{m,n}(x) = 2^{-m/2}\, \psi(2^{-m}x - n)$$

constitute an orthonormal basis for $L^2(\mathbb{R})$. (Other choices for a_0 are possible, but we shall restrict ourselves to $a_0 = 2$ here.) The oldest example of a function ψ for which the $\psi_{m,n}$ defined by (2.6) constitute an orthonormal basis for $L^2(\mathbb{R})$ is the Haar function,

$$\psi(x) = \begin{cases} 1 & 0 \le x < 1/2 \\ -1 & 1/2 \le x < 1 \\ 0 & \text{otherwise}. \end{cases}$$

The Haar basis has been known since Haar (1910). Note that the Haar function does not have good time-frequency localization: its Fourier transform $\hat{\psi}(\xi)$ decays like $|\xi|^{-1}$ for $\xi \to \infty$. Nevertheless we shall use it here for illustration purposes. What follows is a proof that the Haar family does indeed constitute an orthonormal basis. This proof is different from the one in most textbooks; in fact it will use multiresolution analysis as a tool.

In order to prove that the $\psi_{m,n}(x)$ constitute an orthonormal basis, we need to establish that

(1) the $\psi_{m,n}$ are orthonormal
(2) any L^2-function f can be approximated, up to arbitrarily small precision, by a finite linear combination of the $\psi_{m,n}$.

Orthonormality is easy to establish. Since support$(\psi_{m,n}) = [2^m n, 2^m(n+1)]$, it follows that two Haar wavelets of the same scale (same value of m) never overlap, so that $\langle \psi_{m,n}, \psi_{m,n'} \rangle = \delta_{n,n'}$. Overlapping supports are possible if the two wavelets have different sizes, as in Figure 5. It is easy to check, however, that if $m < m'$, then support$(\psi_{m,n})$ lies wholly within a region where $\psi_{m',n'}$ is constant (as on the figure). It follows that the inner product of $\psi_{m,n}$ and $\psi_{m',n'}$ is then proportional to the integral of ψ itself, which is zero.

We concentrate now on how well an arbitrary function f can be approximated by linear combinations of Haar wavelets. Any f in $L^2(\mathbb{R})$ can be arbitrarily well approximated by a function with compact support which is piecewise constant on the $[\ell 2^{-j}, (\ell + 1)2^{-j}[$ (it suffices to take the support and j large enough). We can therefore restrict ourselves to such piecewise constant functions only:

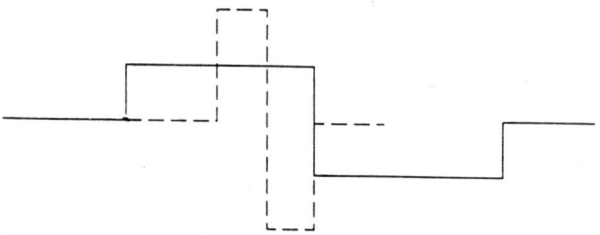

FIGURE 5. Two Haar wavelets; the support of the "narrower" wavelet is completely contained in an interval where the "wider" wavelet is constant.

assume f to be supported on $[-2^{J_1}, 2^{J_1}]$, and to be piecewise constant on the $[\ell 2^{-J_0}, (\ell+1)2^{-J_0}[$, where J_1 and J_0 can both be arbitrarily large (see Figure 6). Let us denote the constant value of $f^0 = f$ on $[\ell 2^{-J_0}, (\ell+1)2^{-J_0}[$ by f^0_ℓ. We now represent f^0 as a sum of two pieces, $f^0 = f^1 + \delta^1$, where f^1 is an approximation to f^0 which is piecewise constant over intervals twice as large as originally, i.e. $f^1|_{[k2^{-J_0+1}, (k+1)2^{-J_0+1}[} \equiv constant = f^1_k$. The values f^1_k are given by the averages of the two corresponding constant values for f^0, $f^1_k = \frac{1}{2}(f^0_{2k} + f^0_{2k+1})$ (see Figure 6). The function δ^1 is piecewise constant with the same stepwidth as f^0; one immediately has

$$\delta^1_{2\ell} = f^0_{2\ell} - f^1_\ell = \frac{1}{2}(f^0_{2\ell} - f^0_{2\ell+1})$$

and

$$\delta^1_{2\ell+1} = f^0_{2\ell+1} - f^1_\ell = \frac{1}{2}(f^0_{2\ell+1} - f^0_{2\ell}) = -\delta^1_{2\ell} \ .$$

It follows that δ^1 is a linear combination of scaled and translated Haar functions:

$$\delta^1 = \sum_{\ell=-2^{J_1+J_0-1}+1}^{2^{J_1+J_0-1}} \delta^1_{2\ell} \psi(2^{J_0-1}x - \ell) \ .$$

We have therefore written f as

$$f = f^0 = f^1 + \sum_\ell c_{-J_0+1,\ell} \ \psi_{-J_0+1,\ell} \ ,$$

where f^1 is of the same type as f^0, but with stepwidth twice as large. We can apply the same trick to f^1, so that

$$f^1 = f^2 + \sum_\ell c_{-J_0+2,\ell} \ \psi_{-J_0+2,\ell} \ ,$$

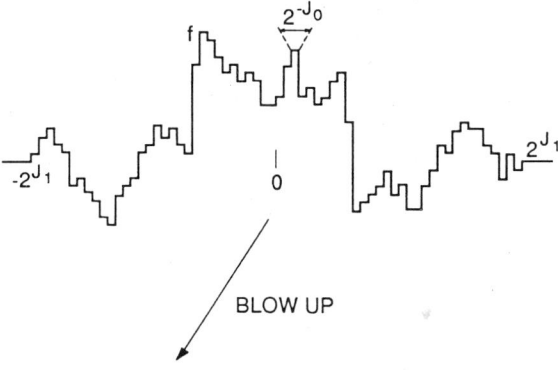

FIGURE 6. (a) A function f with support $[-2^{J_1}, 2^{J_1}]$, piecewise constant on the $[k2^{-J_0}, (k+1)2^{-J_0}[$. (b) A blowup of a portion of f. On every pair of intervals, f is replaced by its average ($\longrightarrow$ f^1); the difference between f and f^1 is δ^1, a linear combination of Haar wavelets.

with f^2 still supported on $[-2^{J_1}, 2^{J_1}]$, but piecewise constant on the even larger intervals $[k2^{-J_0+2}, (k+1)2^{-J_0+2}[$. We can keep going like this, until we have

$$f = f^{J_0+J_1} + \sum_{m=-J_0+1}^{J_1} \sum_{\ell} c_{m,\ell}\, \psi_{m,\ell}\,.$$

Here $f^{J_0+J_1}$ consists of two constant pieces (see Figure 7), with $f^{J_0+J_1}|_{[0,2^{J_1}[} \equiv f_0^{J_0+J_1}$ equal to the average of f over $[0, 2^{J_1}[$, and $f^{J_0+J_1}|_{[-2^{J_1},0[} \equiv f_{-1}^{J_0+J_1}$ the average of f over $[-2^{J_1}, 0[$.

Even though we have "filled out" the whole support of f, we can still keep going with our averaging trick: nothing stops us from widening our horizon from 2^{J_1} to 2^{J_1+1}, and writing $f^{J_1+J_2} = f^{J_1+J_2+1} + \delta^{J_1+J_2+1}$, where

$$f^{J_1+J_2+1}|_{[0,2^{J_1+1}[} \equiv \frac{1}{2} f_0^{J_1+J_2},\ \ f^{J_1+J_2+1}|_{[-2^{J_1+1},0[} \equiv \frac{1}{2} f_{-1}^{J_1+J_2}$$

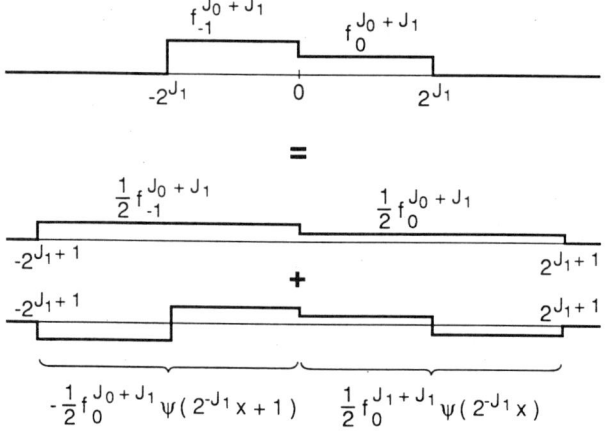

FIGURE 7. The averages of f on $[0, 2^{J_1}]$ and $[-2^{J_1}, 0]$ can be "smeared" out over the bigger intervals $[0, 2^{J_1+1}], [-2^{J_1+1}, 0]$; the difference is a linear combination of very stretched out Haar functions.

and

$$\delta^{J_1+J_2} = \frac{1}{2} f_0^{J_1+J_2} \psi(2^{-J_1-1}x) - \frac{1}{2} f_{-1}^{J_1+J_2} \psi(2^{-J_1-1}x + 1)$$

(see Figure 7). This can again be repeated, leading to

$$f = f^{J_0+J_1+K} + \sum_{m=-J_0+1}^{J_1+K} \sum_{\ell} c_{m,\ell}\, \psi_{m,\ell}\,,$$

where $\text{support}(f^{J_0+J_1+K}) = [-2^{J_1+K}, 2^{J_1+K}]$, and

$$f^{J_0+J_1+K}\big|_{[0,2^{J_1+K}[} = 2^{-K} f_0^{J_0+J_1}, \; f^{J_0+J_1+K}\big|_{[-2^{J_1+K},0[} = 2^{-K} f_{-1}^{J_0+J_1}\,.$$

It follows immediately that

$$\left\| f - \sum_{m=-J_0+1}^{J_1+K} \sum_{\ell} c_{m,\ell}\, \psi_{m,\ell} \right\|_{L^2}^2 = \|f^{J_0+J_1+K}\|_{L^2}^2$$

$$= 2^{-K/2} \cdot 2^{J_1/2}\, [|f_0^{J_0+J_1}|^2 + |f_{-1}^{J_0+J_1}|^2]^{1/2}\,,$$

which can be made arbitrarily small by taking sufficiently large K. As claimed, f can therefore be approximated to arbitrary precision by a finite linear combination of Haar wavelets!

The argument we just saw has implicitly used a "multiresolution" approach: we have written successive coarser and coarser approximations to f (the f^j, averaging f over larger and larger intervals), and at every step we have written the difference between the approximation with resolution 2^{j-1}, and the next coarser level, with resolution 2^j, as a linear combination of the $\psi_{j,k}$.

The Haar basis is a "good" basis for $L^p(\mathbb{R})$, $1 < p < \infty$ (i.e. it is an unconditional basis; see §7). It is however not a suitable basis for smoother function spaces, such as the Sobolev spaces. In the next section, we shall see how the multiresolution approach can be made to work for other, smoother wavelet bases, which then are unconditional bases for a much wider range of functional spaces.

3. Multiresolution analysis.

A multiresolution analysis consists of a sequence of successive approximation spaces V_j. More precisely, the closed subspaces V_j satisfy

$$(3.1) \qquad \ldots V_2 \subset V_1 \subset V_0 \subset V_{-1} \subset V_{-2} \subset \ldots$$

with

$$(3.2) \qquad \overline{\bigcup_{j\in\mathbb{Z}} V_j} \;=\; L^2(\mathbb{R}) \,,$$

$$(3.3) \qquad \bigcap_{j\in\mathbb{Z}} V_j \;=\; \{0\} \,.$$

If we denote by P_j the orthogonal projection operator onto V_j, then (3.2) ensures that $\lim_{j\to-\infty} P_j f = f$ for all $f \in L^2(\mathbb{R})$. There exist many ladders of spaces satisfying (3.1)–(3.3) which have nothing to do with "multiresolution"; the multiresolution aspect is a consequence of the additional requirement

$$(3.4) \qquad f \in V_j \iff f(2^j\cdot) \in V_0 \,.$$

That is, all the spaces are scaled versions of the central space V_0. An example of spaces V_j satisfying (3.1)–(3.4) is

$$V_j = \{f \in L^2(\mathbb{R}); \ \forall k \in \mathbb{Z}: \ f|_{[2^j k,\, 2^j(k+1)[} = \text{constant}\} \,.$$

We shall call this example the Haar multiresolution analysis. It corresponds with our argument in §2.3; see also below. Figure 8 shows what the projection of some f on the Haar spaces V_0, V_{-1} might look like. This example also exhibits another feature that we require from a multiresolution analysis: invariance of V_0 under integer translations,

$$(3.5) \qquad f \in V_0 \Rightarrow f(\cdot - n) \in V_0, \text{ for all } n \in \mathbb{Z} \,.$$

Because of (3.4) this implies that if $f \in V_j$, then $f(\cdot - 2^j n) \in V_j$ for all $n \in \mathbb{Z}$. Finally, we require also that there exists $\phi \in V_0$ so that

$$(3.6) \qquad \{\phi_{0,n}; \ n \in \mathbb{Z}\} \text{ is an orthonormal basis in } V_0$$

where, for all $j, n \in \mathbb{Z}$, $\phi_{j,n}(x) = 2^{-j/2}\, \phi(2^{-j}x - n)$. Together, (3.6) and (3.4) imply that $\{\phi_{j,n}; \ n \in \mathbb{Z}\}$ is an orthonormal basis for V_j, for all $j \in \mathbb{Z}$. This last

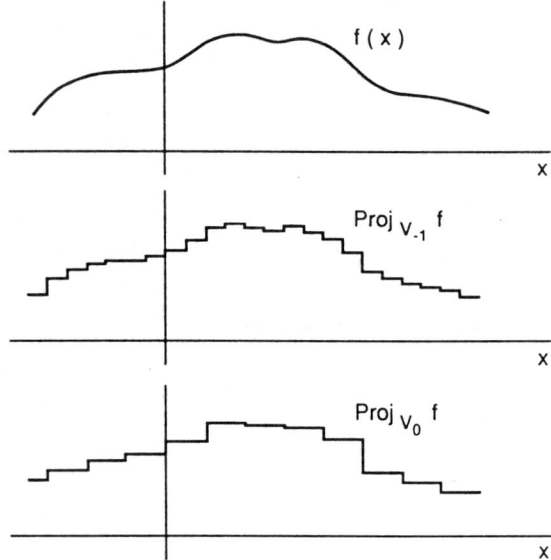

FIGURE 8. A function f and its projections onto V_{-1} and V_0.

requirement (3.6) seems a bit more "contrived" than the other ones; we shall see below that it can be relaxed considerably. In the Haar example, a possible choice for ϕ is the indicator function for $[0,1]$, $\phi(x) = 1$ if $0 \leq x \leq 1$, $\phi(x) = 0$ otherwise. We shall often call ϕ the "scaling function" of the multiresolution analysis.

The basic tenet of multiresolution analysis is that whenever a collection of closed subspaces satisfies (3.1)–(3.6), then there exists an orthonormal wavelet basis $\{\psi_{j,k}; \ j, k \in \mathbb{Z}\}$ of $L^2(\mathbb{R})$, $\psi_{j,k}(x) = 2^{-j/2}\psi(2^{-j}x - k)$, such that, for all f in $L^2(\mathbb{R})$,

$$(3.7) \qquad P_{j-1}f = P_j f + \sum_{k \in \mathbb{Z}} \langle f, \ \psi_{j,k} \rangle \ \psi_{j,k} \ .$$

(P_j is the orthogonal projection onto V_j.) The wavelet ψ can moreover be constructed explicitly. Let us see how.

For every $j \in \mathbb{Z}$, define W_j to be the orthogonal complement of V_j in V_{j-1}. We have

$$(3.8) \qquad V_{j-1} = V_j \oplus W_j \ ,$$

and

$$(3.9) \qquad W_j \perp W_{j'} \ \text{if} \ j \neq j' \ .$$

(If $j > j'$, e.g., then $W_j \subset V_{j'} \perp W_{j'}$.) It follows that, for $j < J$

$$(3.10) \qquad V_j = V_J \oplus \bigoplus_{k=0}^{J-j-1} W_{J-k} \,,$$

where all these subspaces are orthogonal. By virtue of (3.2) and (3.3) this implies

$$(3.11) \qquad L^2(\mathbb{R}) = \bigoplus_{j \in \mathbb{Z}} W_j \,,$$

a decomposition of $L^2(\mathbb{R})$ into mutually orthogonal subspaces. Furthermore, the W_j spaces inherit the scaling property (3.4) from the V_j.

$$(3.12) \qquad f \in W_j \iff f(2^j \cdot) \in W_0 \,.$$

Formula (3.7) is equivalent to saying that, for fixed j, $\{\psi_{j,k}; \ k \in \mathbb{Z}\}$ constitutes an orthonormal basis for W_j. Because of (3.11) and (3.2), (3.3) this then automatically implies that the whole collection $\{\psi_{j,k}; \ j, k \in \mathbb{Z}\}$ is an orthonormal basis for $L^2(\mathbb{R})$. On the other hand, (3.12) ensures that if $\{\psi_{0,k}; \ k \in \mathbb{Z}\}$ is an orthonormal basis for W_0, then $\{\psi_{j,k}; \ k \in \mathbb{Z}\}$ will likewise be an orthonormal basis for W_j, for any $j \in \mathbb{Z}$. Our task thus reduces to finding $\psi \in W_0$ such that the $\psi(\cdot - k)$ constitute an orthonormal basis for W_0.

To construct this ψ, let us write out some interesting properties of ϕ and W_0.

1. Since $\phi \in V_0 \subset V_{-1}$, and the $\phi_{-1,n}$ are an orthonormal basis in V_{-1}, we have

$$(3.13) \qquad \phi = \sum_n h_n \, \phi_{-1,n} \,,$$

with

$$(3.14) \qquad h_n = \langle \phi, \ \phi_{-1,n} \rangle, \text{ and } \sum_{n \in \mathbb{Z}} |h_n|^2 = 1 \,.$$

We can rewrite (3.13) as either

$$(3.15) \qquad \phi(x) = \sqrt{2} \sum_n h_n \, \phi(2x - n)$$

or

$$(3.16) \qquad \hat{\phi}(\xi) = \frac{1}{\sqrt{2}} \sum_n h_n \, e^{-in\xi/2} \, \hat{\phi}(\xi/2) \,,$$

where convergence in either sum holds in L^2-sense. Formula (3.16) can be rewritten as

$$(3.17) \qquad \hat{\phi}(\xi) = m_0(\xi/2) \, \hat{\phi}(\xi/2) \,,$$

where

$$(3.18) \qquad m_0(\xi) = \frac{1}{\sqrt{2}} \sum_n h_n \, e^{-in\xi} \, .$$

Equality in (3.17) holds pointwise almost everywhere. As (3.18) shows, m_0 is a 2π-periodic function in $L^2([0, 2\pi])$.

2. The orthonormality of the $\phi(\cdot - k)$ leads to special properties for m_0. We have

$$\begin{aligned}
\delta_{k,0} &= \int dx \, \phi(x) \, \overline{\phi(x-k)} = \int d\xi \, |\hat{\phi}(\xi)|^2 \, e^{ik\xi} \\
&= \int_0^{2\pi} d\xi \, e^{ik\xi} \sum_{\ell \in \mathbb{Z}} |\hat{\phi}(\xi + 2\pi\ell)|^2 \, ,
\end{aligned}$$

implying

$$(3.19) \qquad \sum_\ell |\hat{\phi}(\xi + 2\pi\ell)|^2 = (2\pi)^{-1} \qquad \text{a.e.}$$

Substituting (3.17) leads to ($\zeta = \xi/2$)

$$\sum_\ell |m_0(\zeta + \pi\ell)|^2 \, |\hat{\phi}(\zeta + \pi\ell)|^2 = (2\pi)^{-1} \, ;$$

splitting the sum into even and odd ℓ, using the periodicity of m_0 and applying (3.19) once more gives

$$(3.20) \qquad |m_0(\zeta)|^2 + |m_0(\zeta + \pi)|^2 = 1 \qquad \text{a.e.}$$

3. Let us now characterize W_0: $f \in W_0$ is equivalent to $f \in V_{-1}$ and $f \perp V_0$. Since $f \in V_{-1}$, we have

$$f = \sum_n f_n \, \phi_{-1,n} \, ,$$

with $f_n = \langle f, \, \phi_{-1,n} \rangle$. This implies

$$(3.21) \qquad \hat{f}(\xi) = \frac{1}{\sqrt{2}} \sum_n f_n \, e^{-in\xi/2} \, \hat{\phi}(\xi/2) = m_f(\xi/2) \, \hat{\phi}(\xi/2) \, ,$$

where

$$(3.22) \qquad m_f(\xi) = \frac{1}{\sqrt{2}} \sum_n f_n \, e^{-in\xi}$$

is a 2π-periodic function in $L^2([0, 2\pi])$; convergence in (3.22) holds pointwise a.e. The constraint $f \perp V_0$ implies $f \perp \phi_{0,k}$ for all k, i.e.

$$\int d\xi \, \hat{f}(\xi) \, \overline{\hat{\phi}(\xi)} \, e^{ik\xi} = 0$$

or

$$\int_0^{2\pi} d\xi \; e^{ik\xi} \sum_\ell \hat{f}(\xi + 2\pi\ell) \; \overline{\hat{\phi}(\xi + 2\pi\ell)} = 0 \; ,$$

hence

(3.23)
$$\sum_\ell \hat{f}(\xi + 2\pi\ell) \; \overline{\hat{\phi}(\xi + 2\pi\ell)} = 0 \; ,$$

where the series in (3.23) converges absolutely in $L^1([-\pi, \pi])$. Substituting (3.17) and (3.21), regrouping the sums for odd and even ℓ (which we are allowed to do, because of the absolute convergence), and using (3.19) leads to

(3.24)
$$m_f(\zeta) \; \overline{m_0(\zeta)} + m_f(\zeta + \pi) \; \overline{m_0(\zeta + \pi)} = 0 \qquad \text{a.e.}$$

Since $\overline{m_0(\zeta)}$ and $\overline{m_0(\zeta + \pi)}$ cannot vanish together on a set of nonzero measure (because of (3.20)), this implies the existence of a 2π-periodic function $\lambda(\zeta)$ so that

(3.25)
$$m_f(\zeta) = \lambda(\zeta) \; \overline{m_0(\zeta + \pi)} \qquad \text{a.e.}$$

and

(3.26)
$$\lambda(\zeta) + \lambda(\zeta + \pi) = 0 \; . \qquad \text{a.e.}$$

This last equation can be recast as

(3.27)
$$\lambda(\zeta) = e^{i\zeta} \; \nu(2\zeta) \; ,$$

where ν is 2π-periodic. Substituting (3.27) and (3.25) into (3.21) gives

(3.28)
$$\hat{f}(\xi) = e^{i\xi/2} \; \overline{m_0(\xi/2 + \pi)} \; \nu(\xi) \; \hat{\phi}(\xi/2) \; ,$$

where ν is 2π-periodic.

4. The general form (3.28) for the Fourier transform of $f \in W_0$ suggests that we take

(3.29)
$$\hat{\psi}(\xi) = e^{i\xi/2} \; \overline{m_0(\xi/2 + \pi)} \; \hat{\phi}(\xi/2)$$

as a candidate for our wavelet. Disregarding convergence questions, (3.28) can indeed be written as

$$\hat{f}(\xi) = \left(\sum_k \nu_k \; e^{-ik\xi} \right) \hat{\psi}(\xi)$$

or

$$f = \sum_k \nu_k \; \psi(\cdot - k) \; ,$$

so that the $\psi(\cdot - n)$ are a good candidate for a basis of W_0. We need to verify that the $\psi_{0,k}$ are indeed an orthonormal basis for W_0. First of all, the properties of m_0 and $\hat{\phi}$ ensure that (3.29) defines indeed an L^2-function $\in V_{-1}$ and $\perp V_0$

(by the analysis above), so that $\psi \in W_0$. Orthonormality of the $\psi_{0,k}$ is easy to check:

$$
\begin{aligned}
\int dx \; \psi(x) \; \overline{\psi(x - k)} &= \int d\xi \; e^{ik\xi} \; |\hat\psi(\xi)|^2 \\
&= \int_0^{2\pi} d\xi \; e^{ik\xi} \sum_\ell |\hat\psi(\xi + 2\pi\ell)|^2 \; .
\end{aligned}
$$

Now

$$
\begin{aligned}
\sum_\ell |\hat\psi(\xi + 2\pi\ell)|^2 &= \sum_\ell |m_0(\xi/2 + \pi\ell + \pi)|^2 \; |\hat\phi(\xi/2 + \pi\ell)|^2 \\
&= \; |m_0(\xi/2 + \pi)|^2 \sum_n |\hat\phi(\xi/2 + 2\pi n)|^2 \\
&\quad + |m_0(\xi/2)|^2 \sum_n |\hat\phi(\xi/2 + \pi + 2\pi n)|^2 \\
&= (2\pi)^{-1} \left[|m_0(\xi/2)|^2 + |m_0(\xi/2 + \pi)|^2 \right] \qquad \text{a.e. (by (3.19))} \\
&= (2\pi)^{-1} \qquad \text{a.e. (by (3.20))} \; .
\end{aligned}
$$

Hence $\int dx \; \psi(x) \; \overline{\psi(x - k)} = \delta_{k0}$. In order to check that the $\psi_{0,k}$ are indeed a basis for all of W_0, it then suffices to check that any $f \in W_0$ can be written as

$$
f = \sum_n \gamma_n \; \psi_{0,n} \; ,
$$

with $\sum_n |\gamma_n|^2 < \infty$, or

$$(3.30) \qquad\qquad\qquad \hat f(\xi) = \gamma(\xi) \; \hat\psi(\xi) \; ,$$

with γ 2π-periodic and $\in L^2([0, 2\pi])$. But this is nothing but (3.28), where it is easy to check that ν is indeed square integrable. We have therefore proved the assertion at the start of this section: there is an orthonormal wavelet basis $\{\psi_{j,k}; \; j, k \in \mathbb{Z}\}$ associated with any multiresolution analysis, and we even have a recipe for the construction of ψ:

$$
\begin{aligned}
(3.31) \qquad\qquad \psi(x) &= \sum_n (-1)^n \; h_{-n+1} \; \phi_{-1,n} \\
&= \sqrt{2} \sum_n (-1)^n \; h_{-n+1} \; \phi(2x - n) \; ,
\end{aligned}
$$

where ϕ is the scaling function of the multiresolution analysis. (Note that (3.31) corresponds to (3.29), except for a change of sign, and a shift by 1 in x, neither of which affect the result.)

Not every orthonormal wavelet basis derives from a multiresolution analysis. There exist "pathological" counterexamples in which ψ has very bad decay. (See Mallat (1989) or Daubechies (1992)). Recently, it was proved in P. Auscher (1992) and in Lemarié-Rieusset (1992) that if ψ has a modicum of decay and smoothness, then it necessarily stems from a multiresolution analysis. Lemarié-Rieusset (1991) contains an earlier proof for compactly supported ψ. More details on these results can also be found in the chapter by P. G. Lemarié-Rieusset in this volume.

To conclude this section, let us see what the recipe (3.31) gives for the Haar multiresolution analysis. In that case $\phi(x) = 1$ for $0 \leq x < 1$, 0 otherwise, hence

$$h_n = \sqrt{2} \int dx \; \phi(x) \; \overline{\phi(2x-n)} = \begin{cases} 1/\sqrt{2} & \text{if } n = 0, 1 \\ 0 & \text{otherwise.} \end{cases}$$

Consequently $\psi = \frac{1}{\sqrt{2}} \phi_{-1,0} - \frac{1}{\sqrt{2}} \phi_{-1,1}$ or

$$\psi(x) = \begin{cases} 1 & \text{if } 0 \leq x < 1/2 \\ -1 & \text{if } 1/2 \leq x < 1 \\ 0 & \text{otherwise,} \end{cases}$$

and we recover indeed the Haar basis.

Of course, the real interest of this formalism lies in the other examples that can be built with it. The whole framework was developed by S. Mallat (1989) and Y. Meyer. The first construction of smooth wavelet bases (Stromberg (1982), which unfortunately went largely unnoticed at the time, Meyer (1985), Lemarié (1988) and Battle (1987)) did not use multiresolution analysis, and seemed much more ad hoc and miraculous. Interestingly enough, Mallat's background in vision analysis played a role in the development of multiresolution analysis (see Daubechies (1988) for a discussion of the connection): an interesting example of feedback from a very applied field to theory.

4. Spline wavelets.

Let us try the constructions in §3 for other multiresolution analysis ladders. One can choose e.g. a ladder of spline spaces, very popular in approximation theory.

$$V_j = \{f \text{ in } L^2(\mathbb{R}); \qquad f \in C^{\ell-1} \text{ and } f \mid_{[2^j k, \; 2^j(k+1)]} \text{ is}$$
$$\text{a polynomial of order } \ell, \text{ for all } k \in \mathbb{Z}\} \; .$$

These are splines of order ℓ, with equispaced knots. The requirements (3.1)–(3.5) are obviously satisfied, but (3.6) is a bit more tricky. The usual B-spline function, i.e. the ℓ-th convolution of ϕ_{Haar} with itself, has the property that it and its integer translates generate all of V_0, but they are not orthonormal. For $\ell = 1$, for instance, we get the tent function $\phi(x) = 1 - |x|$ for $|x| \leq 1$, $\phi(x) = 0$ otherwise, and obviously $\phi(x)$ is not orthogonal to $\phi(x-1)$. This can bé fixed easily however; we can relax (3.6) and replace it by the requirement that the $\phi(x-n)$ constitute a Riesz basis for V_0, i.e. that they span V_0 and that for

$f = \sum_n c_n \phi_{0n} \in V_0$, the norms $\sum_n |c_n|^2$ and $\|f\|^2$ are equivalent, in the sense that

(4.1) $$A|c_n|^2 \leq \left\| \sum_n c_n \phi_{0n} \right\|^2 \leq B \sum_n |c_n|^2 \, ,$$

with $A > 0$, $B < \infty$ and independent of f.

Because

$$\left\| \sum_n c_n \phi_{0n} \right\|^2 = \int d\xi \left| \sum_n c_n e^{in\xi} \right|^2 |\hat{\phi}(\xi)|^2$$

$$= \int_0^{2\pi} d\xi \left| \sum_n c_n e^{in\xi} \right|^2 \sum_{\ell \in \mathbb{Z}} |\hat{\phi}(\xi + 2\pi\ell)|^2 \, ,$$

(4.1) is equivalent with

(4.2) $$0 < \frac{A}{2\pi} \leq \sum_{\ell \in \mathbb{Z}} |\hat{\phi}(\xi + 2\pi\ell)|^2 \leq \frac{B}{2\pi} < \infty \, ,$$

a requirement that is satisfied by the tent function (as well as the higher order B-splines). We can therefore define $\tilde{\phi}$ by

(4.3) $$\widehat{\tilde{\phi}}(\xi) = \frac{\hat{\phi}(\xi)}{\left[2\pi \sum_\ell |\hat{\phi}(\xi + 2\pi\ell)|^2 \right]^{1/2}} \, ;$$

because of the stability conditions (4.2) one easily checks that $\tilde{\phi} \in V_0$, and that the $\tilde{\phi}_{0n}$ span V_0 again, as the ϕ_{0n} did. Moreover $\sum_\ell |\widehat{\tilde{\phi}}(\xi + 2\pi\ell)|^2 = (2\pi)^{-1}$, so that the $\tilde{\phi}_{0n}$ are orthonormal. One can then repeat the recipe of §2:

$$h_n = \langle \tilde{\phi}, \tilde{\phi}_{-1,n} \rangle$$
$$\psi(x) = \sqrt{2} \sum_n (-1)^n h_{-n+1} \tilde{\phi}(2x - n) \, ,$$

and the resulting $\psi_{j,k}$ will constitute an orthonormal basis associated with the given multiresolution analysis. Figures 9 and 10 show the functions $\tilde{\phi}$ and ψ for respectively linear and quadratic splines. These orthonormal spline bases were first constructed, independently and by completely different ad hoc methods by P. G. Lemarié (1989) and G. Battle (1988), before the advent of multiresolution analysis. Note that even though the original B-splines have compact support (of width $\ell + 1$ for splines of order ℓ), the orthogonalization trick (4.3) destroys this property; the resulting $\tilde{\phi}$ and ψ are supported on the whole line (with exponential decay).

The very first orthonormal basis of smooth wavelets, constructed by Stromberg (1982), also consists of spline functions; in terms of multiresolution analysis, the difference with the Battle-Lemarié wavelets is that another choice than (3.29)

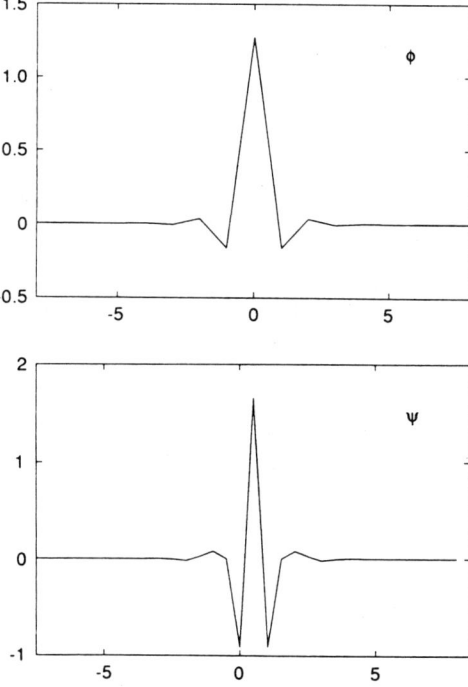

FIGURE 9. Scaling function and wavelet for the linear Battle-Lemarié spline basis.

is made: multiplying (3.29) by any 2π-periodic function of modulus 1 leads to another acceptable candidate for ψ.

Instead of wanting to reduce the spline multiresolution ladder with their very natural but nonorthogonal B-spline basis in every V_j to the case in §3, with orthonormal ϕ_{0n}, one can also try to stick to the B-splines, and characterize W_j, and find ψ, directly.

In this case, one still has

$$\phi(x) = \sum_n c_n \phi(2x - n) \ ,$$

but

$$\int \phi(x) \ \overline{\phi(x - m)} \ dx = \gamma_m \neq \delta_{m0} \ .$$

Note that γ_m can also be written as

$$(4.4) \qquad \gamma_m = \int |\hat{\phi}(\xi)|^2 e^{im\xi} \ d\xi = \int_0^{2\pi} e^{im\xi} \left(\sum_\ell |\hat{\phi}(\xi + 2\pi\ell)|^2 \right) \ d\xi \ .$$

Let us define $c(\xi) = \sum_m c_m e^{-im\xi}$, $\gamma(\xi) = \sum_m \gamma_m e^{-im\xi}$. Because of (4.4),

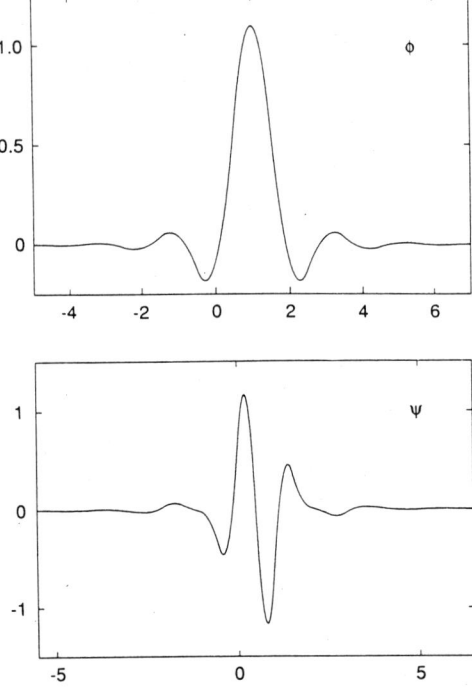

FIGURE 10. Scaling function and wavelet for the quadratic Battle-Lemarié spline basis.

$\gamma(\xi) = \frac{1}{2\pi} \sum_\ell |\hat{\phi}(\xi + 2\pi\ell)|^2$. Saying now that $\psi \in V_{-1} \cap (V_0)^\perp$ amounts to

$$(4.5) \qquad \psi(x) = \sum_n d_n \phi(2x - n) \ ,$$

and

$$(4.6) \qquad 0 = \int \psi(x) \, \overline{\phi(x - k)} = \sum_{n,m} d_n \, c_m \, \gamma_{2k-m+n}, \ \text{for all } k \in \mathbb{Z} \ .$$

With the notation $d(\xi) = \sum_n d_n e^{-in\xi}$, (4.6) can be rewritten as

$$d(\xi) \, \overline{c(\xi)} \, \overline{\gamma(\xi)} + d(\xi + \pi) \, \overline{c(\xi + \pi)} \, \overline{\gamma(\xi + \pi)} = 0 \ ,$$

leading to the candidate solution

$$d(\xi) = -e^{i\xi} \, \overline{c(\xi + \pi)} \, \overline{\gamma(\xi + \pi)} \ ,$$

or

$$(4.7) \qquad d_n = \sum_m (-1)^n c_{-m} \, \gamma_{-n+1+m} \ .$$

Since only finitely many c_n, g_n are nonzero, the same is true for the d_n.

Substituting (4.7) into (4.5) leads to a compactly supported function ψ such that the $\psi(x - k)$ constitute a Riesz basis for W_0; it then easily follows that the $\psi_{j,k}$ constitute a Riesz basis for all of $L^2(\mathbb{R})$. The dual Riesz basis in W_0 of the $\psi(x - k)$ is given by $\tilde{\psi}(x - k)$, with

$$\widehat{\tilde{\psi}}(\xi) = \frac{\hat{\psi}(\xi)}{2\pi \sum_\ell |\hat{\psi}(\xi + 2\ell\pi)|^2} \ ;$$

$\tilde{\psi}$ is not compactly supported, but has exponential decay. The $\tilde{\psi}_{j,k}$ can easily be checked to be the dual Riesz basis for the $\psi_{j,k}$.

This alternative construction was proposed independently by Unser, Aldroubi and Eden (1990) and Chui and Wang (1991); there exist explicit formulas for ψ in terms of spline functions (Chui (1992)). Note that if one chooses to orthonormalize the $\psi(x - k)$, the result is the Battle-Lemarié wavelet again.

It is possible to choose for ψ a compactly supported spline function, and to find another compactly supported function $\tilde{\psi}$ such that the $\psi_{j,k}$, $\tilde{\psi}_{j,k}$ constitute dual Riesz bases. In this case however, one loses the orthonormality between the different j levels (unlike the construction above), and $\tilde{\psi}$ is not a spline function. For details, see Cohen, Daubechies and Feauveau (1992) or Daubechies (1992).

5. Fast algorithms.

Multiresolution analysis leads naturally to a hierarchical and fast scheme for the computation of the wavelet coefficients of a given function. Suppose that we have computed, or are given, the inner products of f with the $\phi_{j,k}$ at some given, fine scale. By rescaling our "units" (or rescaling f) we can assume that the label of this fine scale is $j = 0$. It is then easy to compute the $\langle f, \psi_{j,k} \rangle$ for $j \geq 1$. First of all, we have (see (3.31))

$$\psi = \sum_n g_n \phi_{-1,n} \ ,$$

where $g_n = \langle \psi, \phi_{-1,n} \rangle = (-1)^n h_{-n+1}$. Consequently,

$$
\begin{aligned}
\psi_{j,k}(x) &= 2^{-j/2} \psi(2^{-j}x - k) \\
&= 2^{-j/2} \sum_n g_n 2^{1/2} \phi(2^{-j+1}x - 2k - n) \\
&= \sum_n g_n \phi_{j-1,2k+n}(x) \\
(5.1) \qquad &= \sum_n g_{n-2k} \phi_{j-1,n}(x) \ .
\end{aligned}
$$

(To simplify matters, we assume we are in the orthonormal case. All this can be generalized to the nonorthonormal but dual bases $\psi_{j,k}$, $\tilde{\psi}_{j,k}$ presented at the end of §4.)

It follows that

$$\langle f, \ \psi_{1,k} \rangle \ = \ \sum_n \overline{g_{n-2k}} \ \langle f, \ \phi_{0,n} \rangle \ ,$$

i.e. the $\langle f, \ \psi_{1,k} \rangle$ are obtained by convolving the sequence $(\langle f, \ \phi_{0,n} \rangle)_{n \in \mathbb{Z}}$ with $(\overline{g}_{-n})_{n \in \mathbb{Z}}$, and then retaining only the even samples. Similarly, we have

$$(5.2) \qquad \langle f, \ \psi_{j,k} \rangle \ = \ \sum_n \overline{g_{n-2k}} \ \langle f, \ \phi_{j-1,n} \rangle \ ,$$

which can be used to compute the $\langle f, \ \psi_{j,k} \rangle$ by means of the *same operation* (convolution with $\overline{g}$, decimation by factor 2) from the $\langle f, \ \phi_{j-1,k} \rangle$, if these are known. But, by (3.15),

$$(5.3) \qquad \begin{aligned} \phi_{j,k}(x) \ &= \ 2^{-j/2} \ \phi(2^{-j}x - k) \\ &= \ \sum_n h_{n-2k} \ \phi_{j-1,n}(x) \ , \end{aligned}$$

whence

$$(5.4) \qquad \langle f, \ \phi_{j,k} \rangle \ = \ \sum_n \overline{h_{n-2k}} \ \langle f, \ \phi_{j-1,n} \rangle \ .$$

The procedure to follow is now clear: starting from the $\langle f, \ \phi_{0,n} \rangle$, we compute the $\langle f, \ \psi_{1,k} \rangle$ by (5.2), and the $\langle f, \ \phi_{1,k} \rangle$ by (5.4). We can then apply (5.2), (5.4) again to compute the $\langle f, \ \psi_{2,k} \rangle$, $\langle f, \ \phi_{2,k} \rangle$ from the $\langle f, \ \phi_{1,n} \rangle$, etc. At every step we compute not only the wavelet coefficients $\langle f, \ \psi_{j,k} \rangle$ of the corresponding j-level, but also the $\langle f, \ \phi_{j,k} \rangle$ for the same j-level, which are useful for the computation of the next level wavelet coefficients.

The whole process can also be viewed as the computation of successively coarser approximations of f, together with the difference in "information" between every two successive levels. In this view we start out with a fine-scale approximation to f, $f^0 = P_0 f$ (recall that P_j is the orthogonal projection onto V_j; we shall denote the orthogonal projection onto W_j by Q_j), and we decompose $f^0 \in V_0 = V_1 \oplus W_1$ into $f^0 = f^1 + \delta^1$, where $f^1 = P_1 f^0 = P_1 f$ is the next coarser approximation of f in the multiresolution analysis, and $\delta^1 = f^0 - f^1 = Q_1 f^0 = Q_1 f$ is what is "lost" in the transition $f^0 \rightarrow f^1$. In each of these V_j, W_j spaces we have the orthonormal bases $(\phi_{j,k})_{k \in \mathbb{Z}}$, $(\psi_{j,k})_{k \in \mathbb{Z}}$, respectively, so that

$$f^0 = \sum_n c_n^0 \ \phi_{0,n}, \ f^1 = \sum_n c_n^1 \ \phi_{1,n}, \ \delta^1 = \sum_n d_n^1 \ \psi_{1,n} \ .$$

Formulas (5.2), (5.4) give the effect on the coefficients of the orthogonal basis transformation $(\phi_{0,n})_{n \in \mathbb{Z}} \rightarrow (\phi_{1,n}, \psi_{1,n})_{n \in \mathbb{Z}}$ in V_0:

$$(5.5) \qquad c_k^1 \ = \ \sum_n \overline{h_{n-2k}} \ c_n^0, \ d_k^1 \ = \ \sum_n \overline{g_{n-2k}} \ c_n^0 \ .$$

With the notations $a = (a_n)_{n\in\mathbb{Z}}$, $\bar{a} = (\overline{a_{-n}})_{n\in\mathbb{Z}}$ and $(Ab)_k = \sum_n a_{2k-n} \, b_n$, we can rewrite this as

$$c^1 = \overline{H} \, c^0 \ , \ d^1 = \overline{G} \, c^0 \ .$$

The coarser approximation $f^1 \in V_1 = V_2 \oplus W_2$ can again be decomposed into $f^1 = f^2 + \delta^2$, $f^2 \in V_2$, $\delta^2 \in W_2$, with

$$f^2 = \sum_n c_n^2 \, \phi_{2,n} \ \ \delta^2 = \sum_n d_n^2 \, \psi_{2,n} \ .$$

We have again

$$c^2 = \overline{H} \, c^1 , \ d^2 = \overline{G} \, c^1 \ .$$

Schematically, all this can be represented as in Figure 11. In practice, we will

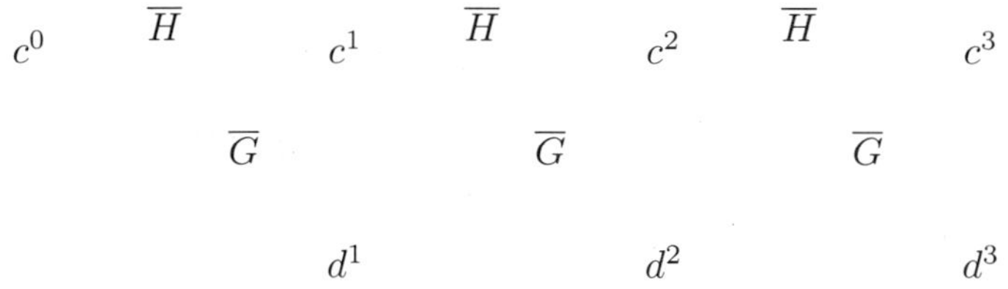

$$c^0 \qquad \overline{H} \qquad c^1 \qquad \overline{H} \qquad c^2 \qquad \overline{H} \qquad c^3$$

$$\overline{G} \qquad\qquad\qquad \overline{G} \qquad\qquad\qquad \overline{G}$$

$$d^1 \qquad\qquad\qquad d^2 \qquad\qquad\qquad d^3$$

FIGURE 11. Schematic representation of (5.5)

stop after a finite number of levels, which means we have rewritten the information in $(\langle f, \phi_{0,n}\rangle)_{n\in\mathbb{Z}} = c^0$ as $d^1, d^2, d^3, \dots, d^J$ and a final coarse approximation c^J, i.e. $(\langle f, \psi_{j,k}\rangle)_{k\in\mathbb{Z}, \, j=1,\dots,J}$ and $(\langle f, \phi_{J,k}\rangle)_{k\in\mathbb{Z}}$. Since all we have done is a succession of orthogonal basis transformations, the inverse operation is given by the adjoint matrices. Explicitly,

$$
\begin{aligned}
f^{j-1} &= f^j + \delta^j \\
&= \sum_k c_k^j \, \phi_{j,k} + \sum_k d_k^j \, \psi_{j,k} \ ,
\end{aligned}
$$

hence

$$
\begin{aligned}
c_n^{j-1} &= \langle f^{j-1}, \phi_{j-1,n}\rangle \\
&= \sum_k c_k^j \, \langle \phi_{j,k}, \phi_{j-1,n}\rangle + \sum_k d_k^j \, \langle \psi_{j,k}, \phi_{j-1,n}\rangle \\
(5.6) \qquad &= \sum_k \left[h_{n-2k} \, c_k^j + g_{n-2k} \, d_k^j \right] \qquad \text{(use (5.1), (5.3))} \ .
\end{aligned}
$$

An important aspect of the whole decomposition is that it is a fast algorithm. Let us go back to the Haar basis for a moment. If we start with N data points c_n^0, then we have to compute $N/2$ averages c_n^1, and $N/2$ differences d_n^1; from the $N/2$ different c_n^1 we compute $N/4$ averages c_n^2 and $N/4$ differences d_n^2, etc. The total number of computations is therefore $2\left(\frac{N}{2} + \frac{N}{4} + \dots\right) = 2N$. For more sophisticated wavelet bases, the "averages" and "differences" involve more than just two numbers, but the same argument holds. If every "generalized average or difference" involves K coefficients of the previous level (rather than

2 as in the Haar case), then the total number of computations is $2KN$ (with KN multiplications, KN additions; this can be reduced further if the h_n have additional structure).

The orthonormal spline bases we saw in §4 have infinitely supported $\tilde{\phi}$ and ψ, resulting in infinitely many nonvanishing h_n. In practice, one needs to truncate to a finite number (otherwise we will hardly have a fast algorithm!). Since $\tilde{\phi}$, and therefore the h_n, have exponential decay, this truncating can in principle be done very easily; in practice one finds that K is rather large. This is one motivation to look at other multiresolution analysis ladders, where the emphasis is on the construction of ϕ associated with a finite number of h_n rather than on the choice of natural spaces V_j.

It should be noted that the fast algorithms associated with an orthonormal wavelet basis are also known, in electrical engineering, as a subband filtering scheme with exact reconstruction. Such schemes were constructed in EE by Smith and Barnwell (1986), Mintzner (1985) and Vetterli (1986), independently of, and in fact before, wavelets.

6. Orthonormal bases of compactly supported wavelet bases.

The easiest way to ensure compact support for the wavelet ψ is to choose the scaling function ϕ with compact support (in its orthogonalized version). It then follows from the definition of the h_n,

$$h_n = \sqrt{2} \int dx \, \phi(x) \, \overline{\phi(2x - n)} \ ,$$

that only finitely many h_n are nonzero, so that ψ reduces to a finite linear combination of compactly supported functions (see (3.31)), and therefore automatically has compact support itself.

For compactly supported ϕ the 2π-periodic function m_0,

$$m_0(\xi) = \frac{1}{\sqrt{2}} \sum_n h_n \, e^{-in\xi} \ ,$$

becomes a trigonometric polynomial. As shown in §4, orthonormality of the $\phi_{0,n}$ implies

(6.1) $$|m_0(\xi)|^2 + |m_0(\xi + \pi)|^2 = 1 \ ,$$

where we have dropped the "almost everywhere" because m_0 is necessarily continuous, so that (6.1) has to hold for all ξ if it holds a.e.

We are also interested in making ψ and ϕ reasonably regular. When we were working with spline spaces, we automatically controlled the regularity of ϕ and ψ. In this different setting, things are not as automatic. First of all, we have a necessary condition:

THEOREM 6.1. Suppose $f \in L^2(\mathbb{R})$ satisfies

$$\langle f_{j,k}, f_{j',k'} \rangle = \delta_{jj'} \delta_{kk'} ,$$

with $f_{j,k}(x) = 2^{-j/2} f(2^{-j}x - k)$. Suppose that f has compact support and that $f \in C^m$, with $f^{(\ell)}$ bounded for $\ell \le m$. Then

(6.2) $$\int dx\, x^\ell \, \tilde{f}(x) = 0 \ for \ \ell = 0, 1, \dots, m .$$

The idea of the proof is very simple. Choose j, k, j', k' so that $f_{j,k}$ is rather spread out, and $f_{j',k'}$ very much concentrated. On the tiny support of $f_{j',k'}$ the slice of $f_{j,k}$ "seen" by $f_{j',k'}$ can be replaced by its Taylor series, with as many terms as are well-defined. Since, however, $\int dx\, \overline{f_{j,k}(x)}\, f_{j',k'}(x) = 0$, this implies that the integral of the product of f and a polynomial of order m is zero. We can then vary the locations of $f_{j',k'}$, as given by k'. For each location the argument can be repeated, leading to a whole family of different polynomials of order m which all give zero integral when multiplied with f. This leads to the desired moment condition. For a true proof, see Daubechies (1992).

Since (see §4) $\hat{\psi}(\xi) = e^{-i\xi/2} \, \overline{m_0(\xi/2 + \pi)} \, \hat{\phi}(\xi/2)$, with $\hat{\phi}(0) = 1$, and since (6.2) is equivalent with $\frac{d^\ell}{d\xi^\ell} \hat{\psi} \big|_{\xi=0} = 0$ for $\ell = 0, 1, \dots, m$, it follows that $\psi \in C^m$ implies that m_0 has a zero of order $m + 1$ in π, or $m_0(\xi) = \left(\frac{1+e^{i\xi}}{2} \right)^{m+1} \mathcal{L}(\xi)$, with $\mathcal{L}$ again a trigonometric polynomial.

In addition to (6.1), we therefore also impose

(6.3) $$m_0(\xi) = \left(\frac{1 + e^{i\xi}}{2} \right)^N \mathcal{L}(\xi) ,$$

for some $N > 1$.

A first question is whether such m_0 exist. Taking the modulus square of (6.3) gives

$$|m_0(\xi)|^2 = \left(\cos^2 \frac{\xi}{2} \right)^N \, |\mathcal{L}(\xi)|^2 ,$$

where $|\mathcal{L}(\xi)|^2$ is a polynomial in $\cos \xi$, which can therefore also be written as a polynomial in $\sin^2 \frac{\xi}{2}$, i.e.

$$|m_0(\xi)|^2 = \left(\cos^2 \frac{\xi}{2} \right)^N P \left(\sin^2 \frac{\xi}{2} \right) ,$$

with P a polynomial. Substituting this into (6.1) leads to an equation for P,

(6.4) $$x^N P(1 - x) + (1 - x)^N P(x) = 1 .$$

Because x^N and $(1 - x)^N$ are two polynomials of degree N which are relatively prime, Bezout's theorem tells us that there exists a unique polynomial P of degree $N - 1$ which solves (6.4). An explicit expression for P is given by

$$P(x) = \sum_{k=0}^{N-1} \binom{N-1+k}{k} x^k \ ,$$

which fortunately is positive for $0 < x < 1$, so that $P\left(\sin^2 \frac{\xi}{2}\right)$ is at least a possible candidate for $|\mathcal{L}(\xi)|^2$. There also exist higher degree solutions P to (6.4); they can be written as

$$P(x) = \sum_{k=0}^{N-1} \binom{N-1+k}{k} x^k + x^N R\left(x - \frac{1}{2}\right)$$

where R is an odd polynomial. We shall restrict ourselves to the lowest degree solution here.

Now that we have a candidate for $|\mathcal{L}(\xi)|^2$, the next question is to find $\mathcal{L}(\xi)$ itself. This can be achieved by the following lemma of Riesz, also known as "spectral factorization",

LEMMA 6.2. Let A be a positive trigonometric polynomial invariant under the substitution $\xi \to -\xi$; A is necessarily of the form

$$A(\xi) = \sum_{m=0}^{M} a_m \ \cos m\xi, \text{ with } a_m \in \mathbb{R} \ .$$

Then there exists a trigonometric polynomial B of order M, i.e.

$$B(\xi) = \sum_{m=0}^{M} b_m \ e^{im\xi}, \text{ with } b_m \in \mathbb{R} \ ,$$

such that $|B(\xi)|^2 = A(\xi)$.

The proof (which we skip here; details for this derivation can be found in many textbooks; they are also given in Daubechies (1988) or Daubechies (1992)) is constructive, so that we have a recipe for $\mathcal{L}(\xi)$ from $P(x)$.

All this leads us to a family of candidates $m_{0,N}$, with N the order of the zero at π, as in (6.2). Next we need to see how this determines ϕ and ψ. This is easy: since we expect $\phi \in L^1$, with $\int \phi(x) dx = 1$, $\hat{\phi}$ is continuous, with $\hat{\phi}(0) = \frac{1}{\sqrt{2\pi}}$, so that $\hat{\phi}(\xi) = m_0(\xi/2) \ \hat{\phi}(\xi/2)$ can be iterated, leading to

$$
\begin{aligned}
\hat{\phi}(\xi) &= \lim_{J \to \infty} \left[\prod_{j=1}^{J} m_0(2^{-j}\xi) \right] \hat{\phi}(2^{-J}\xi) \\
(6.5) \qquad &= (2\pi)^{-1/2} \prod_{j=1}^{\infty} m_0(2^{-j}\xi) \ ,
\end{aligned}
$$

where the infinite product converges because m_0 is a trigonometric polynomial with $m_0(0) = 1$, so that

$$|m_0(\xi) - 1| \leq C|\xi| \text{ for } |\xi| \leq 1 .$$

It is rather straightforward to show (for details see Daubechies (1992)) that the infinite product (6.5) is an entire function of exponential type; more precisely, if

$$m_0(\xi) = \sum_{n=n_1}^{n_2} \alpha_n e^{-in\xi} ,$$

then

$$|\hat{\phi}(\xi)| \leq C_1(1 + |\xi|)^{M_1} e^{N_1|\text{Im } \xi|} \quad \text{if } \text{Im } \xi \geq 0$$

$$|\hat{\phi}(\xi)| \leq C_2(1 + |\xi|)^{M_2} e^{N_2|\text{Im } \xi|} \quad \text{if } \text{Im } \xi \leq 0 ,$$

implying that ϕ is a distribution with support in $[N_1, N_2]$.

On the other hand, ϕ is also in L^2. We have indeed

$$
\begin{aligned}
\int |\hat{\phi}(\xi)|^2 d\xi &= \lim_{J \to \infty} \int_{|\xi| \leq 2^J \pi} |\hat{\phi}(\xi)|^2 \, d\xi \\
(6.6) \qquad &\leq \lim_{J \to \infty} (2\pi)^{-1} \int_{|\xi| \leq 2^J \pi} \prod_{j=1}^{J} |m_0(2^{-j}\xi)|^2 \, d\xi
\end{aligned}
$$

$$\text{(because } |m_0| \leq 1 \text{ by (6.1));}$$

now

$$\int_{|\xi| \leq 2^J \pi} \prod_{j=1}^{J} |m_0(2^{-j}\xi)|^2 \, d\xi$$

$$= \int_0^{2^{J+1}\pi} \prod_{j=1}^{J} |m_0(2^{-j}\xi)|^2 \, d\xi \qquad \text{(because of periodicity)}$$

$$= \int_0^{2^J \pi} \left[\prod_{j=1}^{J-1} |m_0(2^{-j}\xi)|^2 \right] \left[|m_0(2^{-J}\xi)|^2 + |m_0(2^{-J}\xi + \pi)|^2 \right] \, d\xi$$

$$= \int_0^{2^J \pi} \prod_{j=1}^{J-1} |m_0(2^{-j}\xi)|^2 = \ldots = \int_0^{4\pi} \left| m_0 \left(\frac{\xi}{2} \right) \right|^2 \, d\xi$$

$$= \int_0^{2\pi} \left[\left| m_0 \left(\frac{\xi}{2} \right) \right|^2 + \left| m_0 \left(\frac{\xi}{2} + \pi \right) \right|^2 \right] \, d\xi = 2\pi ,$$

so that (6.6) implies $\int |\hat{\phi}(\xi)|^2 d\xi \leq 1$. It follows that ϕ, ψ are compactly supported L^2-functions, and things are looking good. There is one tricky step still, however:

all this is not sufficient to ensure that the $\phi(x - n)$ are orthonormal, nor even independent. A counterexample is

$$
\begin{aligned}
m_0(\xi) &= \left(\frac{1 + e^{-i\xi}}{2}\right)(1 - e^{-i\xi} + e^{-2i\xi}) \\
&= \frac{1 + e^{-3i\xi}}{2} = e^{-3i\xi/2}\cos\frac{3\xi}{2} \ .
\end{aligned}
$$

This satisfies (6.1), as well as $m_0(0) = 1$. Substituting it into (6.5) leads to

$$
\hat{\phi}(\xi) = (2\pi)^{-1/2}\,e^{-3i\xi/2}\,\frac{\sin 3\xi/2}{3\xi/2}
$$

or

$$
\phi(x) = \begin{cases} 1/3 & 0 \le x \le 3 \\ 0 & \text{otherwise} \ . \end{cases}
$$

This is not a "good" scaling function: the $\phi_{0,n}(x) = \phi(x-n)$ are not orthonormal, even though m_0 satisfies (6.1). Another way of looking at this is to see that (3.19) is not satisfied:

$$
\sum_\ell |\hat{\phi}(\xi + 2\pi\ell)|^2 = (2\pi)^{-1}\left[\frac{1}{3} + \frac{4}{9}\cos\ \xi + \frac{2}{9}\cos 2\xi\right] \ .
$$

Note that this means that $\sum_\ell |\hat{\phi}(\xi + 2\pi\ell)|^2 = 0$ for $\xi = \frac{2\pi}{3}$, so that even (4.2) is not satisfied: the $\phi_{0,n}$ are not even a Riesz basis for the space they span.

In order to avoid this kind of mishap, we have to impose extra conditions on m_0 to make sure that ϕ generates a true multiresolution analysis. These conditions ensure that

(6.7) $$\sum_\ell |\hat{\phi}(\xi + 2\pi\ell)|^2 = (2\pi)^{-1}$$

for all ξ. It turns out that this is the crucial condition: once (6.7) is satisfied, everything else follows automatically, and the $\psi_{j,k}$ constitute an orthonormal wavelet basis.

There are several ways of formulating necessary and sufficient conditions on m_0 ensuring that (6.7) holds, mostly due to Cohen (1990) and Lawton (1990); a detailed discussion is given in Daubechies (1992; sections 6.2, 6.3). A sufficient (but not necessary) condition implying (6.7) is (Mallat (1989)):

$$
\min_{|\xi|\le\pi/2}\ |m_0(\xi)| > 0 \ .
$$

Since this is satisfied for the $m_{0,N}$ we constructed above, everything is safe: for each N we have functions ϕ_N, ψ_N, of supportwidth $2N - 1$, and the $2^{-j/2}\,\psi_N(2^{-j}x - k)$, $j, k \in \mathbb{Z}$, constitute an orthonormal basis for $L^2(\mathbb{R})$. Figure 12 shows a few examples for $N = 2, 3, 5$.

How smooth are these functions? Clearly they are not as smooth as we might have hoped: even though we have zeros for m_0 at π of order resp. 2, 3, 5, the resulting ϕ are obviously not C^1, C^2 or C^4. Nevertheless they have higher regularity than the Haar basis (which was after all our goal), and their regularity increases with N. In fact, asymptotically, $\phi_N \in C^{\mu N}$ (for large N), with $\mu \simeq$ .2019 (see Daubechies (1992; chapter 7)); ψ_N has the same regularity as ϕ_N.

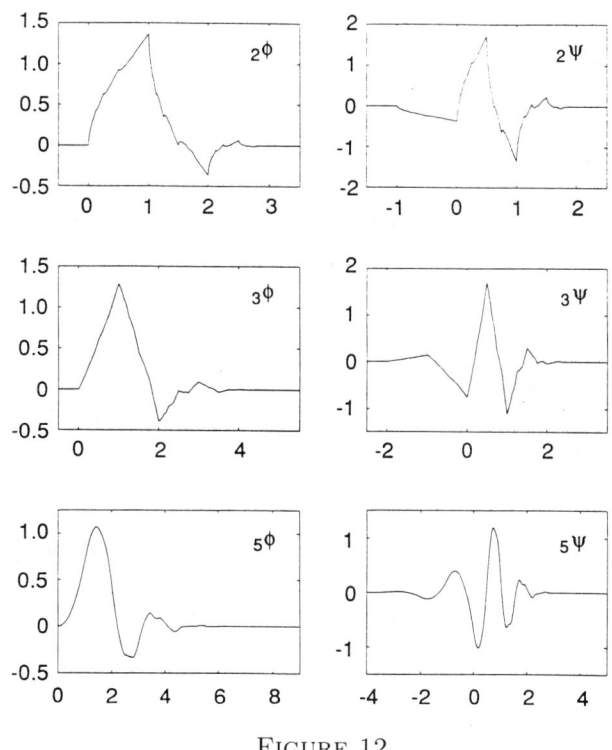

FIGURE 12.

7. Characterization of other function spaces than $\mathbf{L}^2(\mathbb{R})$.

One of the interesting features of smooth wavelet basis is that they provide not only orthonormal bases for $L^2(\mathbb{R})$ but also unconditional bases for many other function spaces.

Let us first review the concept of "unconditional basis". A sequence of vectors $e_1, \ldots, e_n, \ldots$ in a (complex) separable Banach space E is a Schauder basis if, given any $x \in E$, we can find unique $\mu_n \in \mathbb{C}$ so that

$$(7.1) \qquad \lim_{N \to \infty} \left\| x - \sum_{n=1}^{N} \mu_n\, e_n \right\| = 0 \ .$$

The basis is called "unconditional" if in addition, given any sequence $(\mu_n)_{n \in \mathbb{N}}$ in $\mathbb{C}$, there is a criterium, using only the absolute values $|\mu_n|$, to decide whether or not $\sum_{n=1}^{N} \mu_n\, e_n$ converges to some x in E, as $N \to \infty$. Another equivalent way of stating this is the following: whenever $\sum_{n=1}^{N} \mu_n \, e_n$ is in E (in the sense that there exists $x \in E$ so that (7.1) holds), then $\sum_{n=1}^{\infty} \epsilon_n \mu_n e_n \in E$ as well, for any arbitrary choice of the $\epsilon_n = \pm 1$.

The Fourier basis $e_n(x) = e^{2\pi i n x}$, $n \in \mathbb{Z}$, for instance, is an unconditional basis for $L^2([0,1])$ (since it is an orthonormal basis for $L^2([0,1])$, but it is not an unconditional basis for any $L^p([0,1])$ for any $p \neq 1$. One can check for instance that the two series

$$\sum_{n=2}^{\infty} n^{-1/4} \, e^{2\pi i n x} \quad \text{and} \quad \sum_{n=2}^{\infty} n^{-1/4} \, e^{i\sqrt{n}} \, e^{2\pi i n x}$$

both have their worst singularity at $x = 0$; the first one behaves like $|x|^{-3/4}$ for $|x| \to 0$, the second like $|\log x|$ for $x > 0$, $x \to 0$ and like $|x|^{-2}$ for $x < 0$, $x \to 0$ (see Zygmund (1968)). The first is therefore in $L^{7/6}([0,1])$, while the second isn't. Yet the absolute values of their coefficients are the same! No such problem in L^p-spaces exists if one uses the Haar basis. Restricting the Haar basis to only $[0,1]$, i.e. taking $\{\psi_{\text{Haar};j,k}; \; j \leq 0, \; 0 \leq k < 2^{|j|}\}$ and adding to this the constant function 1 on $[0,1]$ gives an unconditional basis for all $L^p([0,1])$ spaces with $1 < p < \infty$ (for $p = 1$ or $p = \infty$, $L^p([0,1])$ does not have an unconditional basis). For smoother function spaces, the discontinuous Haar functions are useless.

This is where smooth wavelet bases are useful. Because of their "logarithmic" treatment of the frequency components, similar to what happens in the Littlewood-Paley approach, they are "good" (i.e. unconditional) bases for L^p-spaces. Because they have good decay and smoothness properties, they are also "good" for function spaces with smoothness requirements, such as the Hölder, Sobolev or Besov spaces. There is no time in this lecture to discuss any of this in detail; let me just give a list of how one can characterize $f \in E$ by means of a criterium on only the $|\langle \psi_{j,k}, f \rangle|$, for several function spaces E.

    1. L^p-spaces

$$f \in L^p(\mathbb{R}) \quad \Longleftrightarrow \quad \left[\sum_{j,k} |\langle f, \psi_{j,k} \rangle|^2 \, |\psi_{j,k}(x)|^2 \right]^{1/2} \in L^p(\mathbb{R})$$

$$\Longleftrightarrow \quad \left[\sum_{j,k} |\langle f, \psi_{j,k} \rangle|^2 \, 2^{-j} \chi_{[2^j k, 2^j(k+1)]}(x) \right]^{1/2} \in L^p(\mathbb{R}) \; .$$

    2. Sobolev spaces

$$f \in W^s(\mathbb{R}) \quad = \quad \left\{ f; \; \int (1 + |\xi|^2)^s \, |\hat{f}(\xi)|^2 \, d\xi < \infty \right\}$$

$$\Longleftrightarrow \quad \sum_{j,k} |\langle f, \psi_{j,k} \rangle|^2 \, (1 + 2^{-2js}) < \infty \; .$$

3. Hölder spaces.

For $s = n + \alpha$, with $n \in \mathbb{N}$ and $0 < \alpha < 1$, we define

$$C^s(\mathbb{R}) = \left\{ f \in L^\infty(\mathbb{R}) \cap C^n(\mathbb{R}); \ \sup_{x,h} \frac{|f^{(n)}(x+h) - f^{(n)}(x)|}{|h|^\alpha} < \infty \right\}.$$

If ψ itself is in $C^r(\mathbb{R})$, with $r > s$ (hence the importance of the smoothness of ϕ, ψ!), then

$$f \in C^s(\mathbb{R}) \Longleftrightarrow \begin{cases} |\langle f, \phi_{0,k}\rangle| \leq C & \text{for all } k \in \mathbb{Z} \\ \text{and} \\ |\langle f, \psi_{j,k}\rangle| \leq C\, 2^{j(s+1/2)} & \text{for all } k \in \mathbb{Z}, \text{ all } j \leq 0 \,. \end{cases}$$

Similar characterizations exist for all the Besov spaces (except those corresponding to L^1 or L^∞ conditions), for the Wiener "bump algebra", for the Hardy space H^1 of Stein and Weiss, for BMO, for the Zygmund class, etc. See Meyer (1990) for a thorough discussion.

Another important aspect of wavelet decompositions is that they are *local*. This can be exploited to characterize local smoothness properties of a function. Again, this can be done by looking at only the absolute values $|\langle f, \psi_{j,k}\rangle|$. For numerically stable computations of local Hölder exponents, it is however often more useful to consider redundant wavelet transforms (see Figure 9.2 in Daubechies (1992), and Mallat and Hwang (1992)).

8. Beyond wavelets.

Wavelets and wavelet transforms have proved useful in a variety of applications which exploit their smoothness, their good concentration in space, their scaling properties, and especially the fact that there exist fast algorithms. Some of these applications will be explained in more detail in this short course. In several of these applications, refinements of the constructions above are needed, such as multidimensional wavelet bases (a first construction of multidimensional wavelets is in Lemarié and Meyer (1986); see also Meyer (1992)) or wavelet bases adapted to an interval (Cohen, Daubechies and Vial (1992)).

There are of course also many applications where wavelets are not the best time frequency tool. Among these we find situations where the Fourier transform is the ideal tool, but also many cases where something intermediary is needed, with ideally a time-frequency analysis adapted to the signal, zooming in on transients (i.e. short-lived high frequency phenomena) with a wavelet-like approach whenever transients are present, but settling for a more Fourier-transform type decomposition for steadily oscillating components. Such more varied approaches can be achieved by means of a generalization of wavelets, called wavelet packets, or by the localized sine transform, an elegant and adaptive variant on the windowed Fourier transform. Both will be discussed in the following chapters.

References

1. E. W. Aslaksen and J. R. Klauder (1968), "Unitary representations of the affine group", *J. Math. Phys.* **9**, pp. 206–211; see also "Continuous representation theory using the affine group", *J. Math. Phys.* **10** (1969), pp. 2267-2275.
2. P. Auscher (1992), "Solution of two problems on wavelets", preprint, submitted to *Annals of Mathematics*.
3. G. Battle (1987), "A block spin construction of ondelettes. Part I: Lemarié functions", *Comm. Math. Phys.* **110**, pp. 601–615.
4. A. P. Calderón (1964), "Intermediate spaces and interpolation, the complex method", *Studia Math.* **24**, pp. 113–190.
5. C. K. Chui (1992), *An Introduction to Wavelets*, Academic Press, New York.
6. C. K. Chui and J. Z. Wang (1991), "A cardinal spline approach to wavelets", *Proc. Amer. Math. Soc.* **113**, pp. 785–793 and "On compactly supported spline wavelets and a duality principle", *Trans. Amer. Math. Soc.* **330** (1992) pp. 903–915.
7. A. Cohen (1990), "Ondelettes, analyses multirésolutions et filtres miroir en quadrature", *Ann. Inst. H. Poincaré, Analyse non linéaire* **7**, pp. 439–459.
8. A. Cohen, I. Daubechies and J. C. Feauveau (1990), "Biorthogonal bases of compactly supported wavelets", *Comm. Pure Appl. Math.* **45**, pp. 485–500.
9. A. Cohen, I. Daubechies and P. Vial (1992), "Wavelets on the interval and fast wavelet transforms", preprint, submitted to *Applied and Computational Harmonic Analysis*.
10. I. Daubechies (1988), "Time-frequency localization operators: a geometric phase space approach," *IEEE Trans. Inf. Th.* **34**, pp. 605–612.
11. I. Daubechies (1992), *Ten lectures on wavelets*, CBMS Lecture Notes nr. 61., SIAM, Philadelphia.
12. N. Delprat, B. Escudié, P. Guillemain, R. Kronland-Martinet, Ph. Tchamitchian and B. Torrésani (1992), "Asymptotic wavelet and Gabor analysis: extraction of instantaneous frequencies", *IEEE Trans. Inf. Th.* **38**, pp. 644–664.
13. M. Frazier and B. Jawerth (1988), "The ϕ-transform and applications to distribution spaces", in *Function Spaces and Application*, M. Cwikel et al., eds., Lecture Notes in Mathematics 1302, pp. 233–246 (Springer, Berlin); see also "A discrete transform and decompositions of distribution spaces", *J. Func. Anal.*, (1990).
14. M. Holschneider and Ph. Tchamitchian (1990), "Régularité locale de la fonction 'non-différentiable' de Riemann", pp. 102–124 in Lemarié (1990).
15. J. R. Klauder and B.-S. Skagerstam (1985), *Coherent states*, (World Scientific, Signapore).
16. W. Lawton (1990), "Tight frames of compactly supported wavelets", *J. Math. Phys.* **31**, pp. 1898–1901; see also "Necessary and sufficient conditions for constructing orthonormal wavelet bases", *J. Math. Phys.* **32**, (1991) pp. 57–61.
17. P. G. Lemarié (1988), "Une nouvelle base d'ondelettes de $L^2(\mathbb{R}^n)$", *J. de Math. Pures et Appl.* **67**, pp. 227–236.
18. P. G. Lemarié and Y. Meyer (1986), "Ondelettes et bases hilbertiennes", *Rev. Mat. Iberoamericana* **2**, pp. 1–18.
19. P. G. Lemarié-Rieusset (1991), "Sur l'existence des analyses multirésolutions en
20. P. G. Lemarié and Y. Meyer (1986), "Ondelettes et bases hilbertiennes", *Rev. Mat. Iberoamericana* **2**, pp. 1–18. théorie des ondelettes", preprint, to appear in *Rev. Mat. Iberoamericana*.
21. P. G. Lemarié-Rieusset (1992), "Projecteurs invariants, matrice de dilation, ondelettes et analyses multirésolutions", preprint, submitted to *Rev. Mat. Iberoamericana*.
22. S. Mallat (1989), "Multiresolution approximation and wavelets", *Trans. Am. Math. Soc.* **315**, pp. 69-88.
23. S. Mallat and S. Zhong (1992), "Characterization of signals from multiscale edges," *IEEE Trans. PAMI* **14**.
24. Y. Meyer (1985), "Principe d'incertitude, bases hilbertiennes et algèbres d'opérateurs", Séminaire Bourbaki, 1985-1986, nr. 662.
25. Y. Meyer (1990), *Ondelettes ét opérateurs, I: Ondelettes, II: Opérateurs de Calderón-Zygmund, III: Opérateurs multilinéaires*, Hermann, Paris. An English translation is being

prepared by the Cambridge University Press 1992.

26. F. Mintzer (1985), "Filters for distortion-free two-band multirate filter banks", *IEEE Trans. Acoust. Speech Signal Process.* **33**, pp. 626-630.

27. M. J. T. Smith and T. P. Barnwell III (1986), "Exact reconstruction techniques for tree-structured subband coders", *IEEE Trans. ASSP* **34**, pp. 434–441; the basic results were already presented at the *IEEE Int. Conf. ASSP*, March 1984, San Diego.

28. J. O. Stromberg (1982), "A modified Franklin system and higher order spline systems on $\mathbb{R}^n$ as unconditional bases for Hardy spaces", Conf. in honor of A. Zygmund, Vol. II, pp. 475–493, ed. W. Beckner et al. (Wadsworth math. series).

29. M. Unser, A. Aldroubi and M. Eden (1990), "A family of polynomial spline wavelet transform", NCRR report 153/90, Nat. Inst. Health.

30. M. Vetterli (1986), "Filter banks allowing perfect reconstruction", *Signal Proc.* **10**, pp. 219–244; these results were already presented as "Splitting a signal into subsampled channels allowing perfect reconstruction", *IASTED Conf. on Applied Sig. Proc. and Dig. Filt.*, June 1985, Paris.

31. A. Zygmund (1968), *Trigonometric Series*, 2nd ed., Cambridge University Press, Cambridge.

AT&T BELL LABORATORIES, 600 MOUNTAIN AVENUE, MURRAY HILL, NEW JERSEY, 07974

E-mail: ingrid@research.att.com

Proceedings of Symposia in Applied Mathematics
Volume **47**, 1993

Wavelets and Operators

YVES MEYER

1. Introduction.

Wavelet analysis is a remarkable tool for analyzing functions of one or several variables that appear in mathematics or in signal and image processing. In these notes we address the possibility of using wavelet analysis for studying operators. The operator under study is denoted by T and is acting on some Hilbert space H. Our first problem will be to find a wavelet based criterion for the continuity of T on H as well as an estimate for the operator norm of T. Our second problem will be to compute $f(T)$ whenever it has a meaning.

These two problems have trivial answers if one can find an orthonormal basis for H in which T is diagonalized. But this situation is exceptional and a more realistic approach will be to find an unconditional basis or a frame (these concepts will be defined below) in which our operator has an almost diagonal representation.

This approach is rather new. We demand that the basis (1) should be adapted to our operator T, and (2) should be accessed through efficient algorithms. This means that we will prefer a basis with good algorithmic properties in which our operator has a sparse representation to a non accessible basis in which this operator would eventually be diagonal.

Before entering into a more detailed discussion let us explain our methodology with two interesting quotations. The first is extracted from D. Marr's book [**20**]:

> "A representation is a formal system for making explicit certain entities or types of information, together with a specification of how the system does this... For example, the Arabic, Roman and binary numerical systems are all formal systems for representing numbers... A musical score provides a way of representing a symphony...

1991 *Mathematics Subject Classification.* Primary 35A27, 35S05, 44A15.

How information is presented can greatly affect how easy it is to do different things with it. This is evident even from our numbers example: it is easy to add, to subtract and even to multiply if the Arabic or binary representations are used, but it is not all easy to do these things – especially multiplication – with Roman numerals. This is a key reason why the Roman culture failed to develop mathematics in the way the earlier Arabic culture had..."

Let us immediately provide the reader with an example taken from mathematics.

If one is given the Fourier series expansion of a function, it is difficult to relate properties like differentiability at given points to estimates on the coefficients. However this problem is trivial if one is given an orthonormal wavelet expansion of this function. A spectacular example is given by the celebrated Riemann function $\sum_1^\infty n^{-2} \sin(n^2 x)$; for more details, the reader is referred to [23].

A second quotation is extracted from [11].

"Many problems in analysis have natural formulations as questions of continuity of linear operators defined on spaces of functions or distributions. Such questions can often be answered by relatively straightforward techniques if they can first be reduced to the study of the operator on an appropriate class of simple elements which, in some convenient sense, generate the entire space... The selection of an orthonormal basis which diagonalizes or nearly diagonalizes an operator... offers us a similar type of approach."

Our task will be to try to understand the subtle relations existing between operator theory, wavelets and partitioning of the "time-frequency plane". For what type of operators should one prefer "time-scale wavelets" to "time-frequency wavelets"? Should new wavelets be tailored?

Before wavelets existed, the best tool for analyzing an operator was the so-called Cotlar lemma. For the convenience of the reader we give it in full detail and list some of its applications.

With the advent of wavelets, a second class of tools became possible, using Schur's lemma. In order to apply Schur's lemma, one needs to find a basis (or a frame) in which our operator T has an almost diagonal representation. The flexibility of the method relies on being provided with large libraries of frames which are adapted to various situations. This is where I. Daubechies' book [14] can be consulted.

After scrutinizing some examples, we will be led to yet a third methodology which is the "non standard representation" of an operator. This methodology is the only existing tool which works for an important class of operators defined by A. Calderón. We will conclude with a recent result concerning the "non standard representation".

2. A problem raised by A. Zygmund.

A. Zygmund often asked the following question: is it possible to prove that the Hilbert transform is bounded on $L^2(\mathbb{R})$ without using the Plancherel formula? The Hilbert transform H is defined as the convolution with the distribution $S = \frac{1}{\pi}$ p.v. $\left(\frac{1}{x}\right)$. The Fourier transform of S is $\hat{S}(\xi) = -i$ sign (ξ). Therefore H is unitary. This approach is strictly limited to convolution operators T which, in some sense, are "diagonal" in the "Fourier basis". A. Zygmund was looking for more general tools which could work as well on operators of the form

$$(2.1) \qquad\qquad Tf(x) \;=\; \text{p.v.} \int K(x,y)\, f(y)\, dy$$

where $K(x,y)$ belongs to some general class of "singular kernels".

Loosely speaking, a singular integral operator is defined by a representation (2.1) which has the property that the corresponding non negative operator $\int |K(x,y)|\,|f(y)|\,dy$ is unbounded. In the case of the Hilbert transform, the corresponding non negative operator is not defined since the integral diverges on the support of f.

Studying a singular integral operator means understanding the delicate cancellations that provide boundedness. A. Calderón was interested in kernels satisfying the three following properties:

$$(2.2) \qquad\qquad |K(x,y)| \;\leq\; C_0|x-y|^{-1}$$

$$(2.3) \qquad\qquad \left|\frac{\partial}{\partial x} K(x,y)\right| \;\leq\; C_1|x-y|^2$$

$$(2.4) \qquad\qquad K(x,y) \;=\; -K(y,x)\,.$$

If we add $K(x,y) = S(x-y)$, which means that T is a convolution operator, these properties imply that T is bounded on $L^2(\mathbb{R})$.

What happens in the general case? How should one attack these operators?

M. Cotlar provided us with an interesting approach to such problems.

3. Cotlar's lemma.

A first answer to A. Zygmund's problem was given by M. Cotlar ([**12**], p. 145).

Let us consider a Hilbert space H and let $T_j : H \to H$ be a collection of bounded linear operators acting on H. The index set J is either finite or infinite but does not have any specific structure. The adjoint of T_j is denoted by T_j^* and we introduce the numbers

$$(3.1) \qquad\qquad \omega(j,k) \;=\; \|T_j^* T_k\|\,, \quad j \in J\,,\; k \in J$$

and

$$(3.2) \qquad\qquad \tilde{\omega}(j,k) \;=\; \|T_j T_k^*\|\;\;, \quad j \in J\,,\; k \in J\,.$$

Here and in what follows, $\| \cdot \|$ denotes the (usual) operator norm. Observe that $\omega(j,k) = \omega(k,j)$ and similarly for $\tilde{\omega}$.

With these notations, one has

THEOREM 3.1. *Let us assume that there exist two constants C_0 and C_1 such that*

$$(3.3) \qquad\qquad \sup_{j \in J} \sum_{k \in J} \sqrt{\omega(j,k)} \leq C_0$$

and

$$(3.4) \qquad\qquad \sup_{j \in J} \sum_{k \in J} \sqrt{\tilde{\omega}(j,k)} \leq C_1 \, .$$

Then one has

$$(3.5) \qquad\qquad \left\| \sum_{j \in J} T_j \right\| \leq \sqrt{C_0 \, C_1}$$

and for each $x \in H$, the series $\sum_{j \in J} T_j(x)$ is norm convergent.

In many examples, $J = \mathbf{Z}$ and both $\omega(j,k)$ and $\tilde{\omega}(j,k)$ can be bounded by $C \exp(-\varepsilon|j-k|)$ for some positive ε. Then Cotlar's lemma applies. It is interesting to observe that we have $\|T_j\|^2 = \|T_j^* T_j\| \leq C$ in such a case and that $\sum \|T_j\|$ may diverge (and often does).

In Cotlar's lemma, (3.3) and (3.4) are not affected by any permutation on the index set J. This implies that the convergence of the series $\sum_{j \in J} T_j(x)$ is not altered by such permutations. This observation leads to the following definition.

DEFINITION 3.2. *Let B be a Banach space and x_j, $j \in J$, be a collection of vectors in B. The series $\sum_{i \in J} x_j$ is summable if for every $\varepsilon > 0$, there exists a finite subset $F \subset J$ such that, for any F' which is finite and disjoint from F, one has*

$$\left\| \sum_{j \in F'} x_j \right\| \leq \varepsilon \, .$$

Then $\sum_{j \in J} x_j = x$ belongs to B and the convergence of this series is not altered if one replaces x_j by $\lambda_j x_j$ where $|\lambda_j| \leq 1$ (the sum of the series obviously changes).

Let us provide the reader with some examples.

A series $\sum_0^\infty x_k$ of real (or complex) numbers is summable if and only if $\sum_0^\infty |x_k| < \infty$; the same is true if the x_k belong to a finite dimensional vector space. If B happens to be a Hilbert space and if $\sum_{j \in J} x_j$ is summable, then one has $\sum \|x_j\|^2 < \infty$ but the converse is not true.

Finally whenever Cotlar's lemma applies to the series $\sum_{j \in J} T_j$, then for every $x \in H$, $\sum_{j \in J} T_j(x)$ is a summable series.

4. Applications of Cotlar's lemma.

Let us return to the problem raised by A. Zygmund and ask for a proof for the boundedness of the Hilbert transform without Plancherel's formula.

We define $\theta_j(x) = 2^j \theta(2^j x)$ where $\theta(x) = \frac{1}{\pi x}$ when $1 \leq |x| \leq 2$ and 0 elsewhere. We denote by T_j the convolution operator with θ_j and observe that

$$(4.1) \qquad\qquad H = \sum_{-\infty}^{\infty} T_j \,.$$

Then $T_k^* T_j$ and $T_k T_j^*$ are given by the convolution with $\theta_j * \theta_k$ (if we ignore irrelevant signs). But a brute force computation yields $\|\theta_j * \theta_k\|_1 \leq C \, 2^{-|j-k|}$. Therefore Cotlar's lemma can be applied.

This is due to M. Cotlar. In the first version of Cotlar's lemma one had $T_k^* T_j = T_j T_k^*$ as in the previous example. The second version (i.e. Theorem 3.1) was found independently by M. Cotlar and E. Stein.

A second application came in [18], where Knapp and Stein were studying singular integral operators on nilpotent Lie groups. The Fourier method did not work but Cotlar's lemma was a success story.

A third application was obtained by A. Calderón and R. Vaillancourt [13]. They were interested in pseudo-differential operators. Pseudo-differential operators are "almost diagonal" in the "Fourier basis" and the precise meaning of this sentence is given by the following assumptions:

$$(4.2) \qquad\qquad T\big(e^{ix\cdot\xi}\big) = \sigma(x,\xi)\, e^{ix\cdot\xi}$$

and there exists an exponent $\delta \in [0,1)$ and constants $C_{\alpha,\beta}$, $\alpha \in \mathbb{N}^n$, $\beta \in \mathbb{N}^n$, such that

$$(4.3) \qquad\qquad |\partial_\xi^\alpha \partial_x^\beta \, \sigma(x,\xi)| \leq C_{\alpha,\beta}(1+|\xi|)^{\delta(|\beta|-|\alpha|)}$$

where $|\beta| = \beta_1 + \cdots + \beta_n$ and similarly for $|\alpha|$.

They successfully cut $\sigma(x,\xi)$ into elementary pieces defined by $\sigma(x,\xi)\,\varphi_j(x,\xi)$, $j \in J$. The cut-off functions $\varphi_j(x,\xi)$ belong to $C_0^\infty(\mathbb{R}^n \times \mathbb{R}^n)$ and have the sharpest localization in the phase space which is allowed by (4.3). Then Cotlar's lemma applies to the corresponding operators defined by these elementary symbols $\sigma(x,\xi)\,\varphi_j(x,\xi)$.

The last problem to be discussed in this section is ; what happens when $\delta = 1$?

The corresponding operators T are not bounded on $L^2(\mathbb{R}^n)$ in general. However they still admit a representation of the form $\int K(x,y)\, f(y)\, dy$ where $K(x,y)$ is a singular integral with estimates generalizing (2.2) and (2.3).

5. What is wrong with Cotlar's lemma?

Before answering this question, let us specify how Cotlar's lemma can be used to prove that a singular operator is bounded on L^2.

We need the following definition:

DEFINITION 5.1. An *integral operator* T is a bounded linear operator $T :$ $L^2(\mathbb{R}^n) \to L^2(\mathbb{R}^n)$ such that there exists another bounded linear operator $L :$ $L^2(\mathbb{R}^n) \to L^2(\mathbb{R}^n)$ with the following two properties

(5.1) $f(x) \geq 0$ a.e.   implies $L(f)(x) \geq 0$ a.e.

(5.2) $|Tf(x)| \leq L(|f|)(x)$ for any f in $L^2(\mathbb{R}^n)$.

Such a non-negative operator L is given by

(5.3) $$Lf(x) = \int f(y) \, d\lambda(x, y)$$

where $d\lambda(x, y)$ is a non-negative measure. Condition (5.2) means that the distributional kernel $K(x, y)$ of T is a signed measure; it also implies that an integral operator remains bounded after suppressing all the cancellations that exist in its distributional kernel.

In practical applications, we still need to prove this boundedness. But the tools needed will be much more primitive than those required for dealing with singular kernels. For example it will suffice to use Schur's lemma.

Schur's lemma. Let $K(x, y)$ be a non-negative measurable functions on $\mathbb{R}^n \times \mathbb{R}^n$. Let us assume that there exist two positive measurable function $\omega(x)$ and $\tilde{\omega}(x)$ which are finite everywhere and which satisfy the two following properties

(5.4) $$\int K(x, y) \, \omega(y) \, dy \leq \tilde{\omega}(x) \qquad \text{(a.e.)}$$

(5.5) $$\int K(x, y) \, \tilde{\omega}(x) \, dy \leq \omega(y) \qquad \text{(a.e.)} .$$

Then the integral operator defined by $Tf(x) = \int K(x, y) \, f(y) \, dy$ is bounded on $L^2(\mathbb{R}^n)$ and $\|T\| \leq 1$.

For a proof, the reader is referred to [11], p. 22.

Let us now return to the question: how does one use Cotlar's lemma to prove that a singular integral operator $Tf(x) = \int S(x, y) \, f(y) \, dy$ is bounded on $L^2(\mathbb{R}^n)$?

One needs to find an expansion of the kernel $S(x, y)$ as a series $\sum_{j \in J} S_j(x, y)$ yielding operators $T_j f(x) = \int S_j(x, y) \, f(y) \, dy$ with the following properties

(5.6) T_j is an integral operator for each $j \in J$

(5.7) $$T_k^* T_j \, f(x) = \int A_{j,k}(x, y) \, f(y) \, dy$$

and the corresponding non-negative operator defined by the kernel $|A_{j,k}(x, y)|$ has an operator norm bounded by $\omega(j, k)$

(5.8)   similarly for $T_k T_j^*$

(5.9) $\omega(j,k)$ and $\tilde{\omega}(j,k)$ satisfy (3.3) and (3.4) .

This precise definition means that the cancellations which exist in the distributional kernel $S(x,y)$ are preserved in the building blocks $S_j(x,y)$ and are fully used to obtain sharp estimates on $|A_{j,k}|$.

The main difficulty in this program consists in finding the decomposition $S(x,y) = \sum_{j \in J} S_j(x,y)$.

Let us consider an example due to A. Calderón. Let $A : \mathbb{R}^n \to \mathbb{R}$ be a Lipschitz function which means that $|A(y) - A(x)| \leq C_0|y - x|$ for every $x, y \in \mathbb{R}^n$. Now consider the singular kernel $S(x,y) = $ p.v. $(A(x) - A(y))|x - y|^{-n-1}$. The corresponding operator is bounded on L^2. This theorem has now received more than ten distinct proofs.

However we do not know how Cotlar's lemma can be used to prove this fact; we do not even know if it is possible.

A second drawback in Cotlar's lemma is the following: it is an expensive tool when compared to less abstract methods. For example, if one returns to the pseudo-differential operators with symbols satisfying (4.3), then $n/2 + \varepsilon$ derivatives in x and ξ suffice to yield an L^2 estimate when $\delta = 0$. These derivatives can even be measured in the Sobolev spaces scale. When one applies Cotlar's lemma to this operator class, one needs twice as many derivatives.

We will treat this example with Schur's lemma in the next section.

6. Schur's lemma.

Let us begin with the definition of a frame. A frame for a Hilbert space H is a substitute for an orthonormal basis. This substitute provides us with a useful flexibility in the applications and is not a limitation when Schur's lemma is used.

DEFINITION 6.1. Let H be a Hilbert space. A collection $(e_j)_{j \in J}$ of vectors in H is a frame if the two following properties are satisfied

(6.1)   there exists a constant C_0 such that, for any (finite) sequence
$$(\alpha_j) \in l^2(J) \quad \text{one has} \quad \left\| \sum \alpha_j e_j \right\| \leq C_0 \left(\sum |\alpha_j|^2 \right)^{1/2}$$

(6.2)   the mapping from $l^2(J)$ into H defined by $(\alpha_j) \to \sum \alpha_j e_j$ is onto .

By Banach's theorem (6.2) implies the following property

(6.3)   there exists a constant C_1 such that every
$$x \in H \quad \text{admits a decomposition} \quad x = \sum \alpha_j e_j$$
$$\text{where} \quad \left(\sum |\alpha_j|^2 \right)^{1/2} \leq C_1 \|x\| .$$

Moreover it can be proved that this "cheap expansion" is provided by a linear algorithm: $\alpha_j = \langle x, f_j \rangle$, where $(f_j)_{j \in J}$ is a dual frame.

If the mapping defined by (6.2) happens to be an isomorphism between $l^2(J)$ and H, $(e_j)_{j \in J}$ is called a Riesz basis (or an unconditional basis). If this mapping is a unitary isomorphism, $(e_j)_{j \in J}$ is an orthonormal basis.

Interesting examples will be given later on. Here is a trivial one where $H = \mathbb{R}^2$. We consider the three vectors

$$e_0 = (1,0) , \quad e_1 = \left(-\frac{1}{2}, \frac{\sqrt{3}}{2}\right) \quad \text{and} \quad e_2 = \left(-\frac{1}{2}, -\frac{\sqrt{3}}{2}\right) .$$

Then each $x \in \mathbb{R}^2$ admits the decomposition

$$(6.4) \qquad\qquad x = \frac{2}{3} \left(\langle x, e_0 \rangle e_0 + \langle x, e_1 \rangle + \langle x, e_2 \rangle e_2 \right)$$

and the coefficient $\frac{2}{3}$ measures the redundancy of the frame (three vectors are used for a two dimensional vector space).

Let us see how this fits in with Schur's lemma. We consider a densely defined operator $T : V \to H$, $V \subset H$, and we would like to prove that T is bounded on H. Assume that we are given a frame $(e_j)_{j \in J}$ in H and that each e_j belongs to V.

One computes the entries

$$\alpha(j,k) = \langle T(e_j), e_k \rangle , \quad j \in J , \ k \in J$$

and considers the matrix $A = (\alpha(j,k))_{(j,k) \in J \times J}$. Since $(e_j)_{j \in J}$ is a frame, proving that T is bounded on H is equivalent to proving that A is bounded on $l^2(J)$. Here is the place where the discrete form of Schur's lemma is used

Schur's lemma. If one can find two sequences $\omega(j)$, $j \in J$, and $\tilde{\omega}(j)$, $j \in J$, of positive real numbers such that

$$(6.5) \qquad\qquad \text{for each } j \in J , \quad \sum_{k \in J} |\alpha(j,k)|\, \omega(k) \leq \tilde{\omega}(j)$$

and

$$(6.6) \qquad\qquad \text{for each } k \in J , \quad \sum_{j \in J} |\alpha(j,k)|\, \tilde{\omega}(j) \leq \omega(k) ,$$

then A is bounded on $l^2(J)$ and $\|A\| \leq 1$.

In many applications $|\alpha(k,k)| \geq \delta > 0$ which implies $\tilde{\omega} \geq \delta\omega \geq \delta^2\tilde{\omega}$ and one then choses $\omega = \tilde{\omega}$.

The proof of Schur's lemma is trivial and runs as follows. Assuming that $\sum |x_j|^2 \leq 1$, we consider $y_j = \sum_k \alpha(j,k)x_k$. Cauchy-Schwarz' inequality yields

$$
\begin{aligned}
\sum_k |\alpha(j,k)|\,|x_k| \quad &\leq \quad \left(\sum_k |\alpha(j,k)| \omega(k) \right)^{1/2} \left(\sum_k |\alpha(j,k)|\, \omega^{-1}(k)\, |x_k|^2 \right)^{1/2} \\
&\leq \quad (\tilde{\omega}(j))^{1/2} \left(\sum_k |\alpha(j,k)|\, \omega^{-1}(k)\, |x_k|^2 \right)^{1/2} .
\end{aligned}
$$

Therefore

$$
\sum_j |y_j|^2 \leq \sum_j \sum_k \tilde{\omega}(j)|\alpha(j,k)|\, \omega^{-1}(k)\, |x_k|^2 \leq \sum_k |x_k|^2 .
$$

Schur's lemma looks very primitive when compared to Cotlar's lemma. But in many examples it yields better results and simpler proofs.

7. Schur's lemma and pseudo-differential operators.

Let us return to the symbol class defined by (4.3) with $\delta = 0$. We will construct a frame which is adapted to the corresponding operators.

We start by constructing window functions. A function $w(x)$ will be a window if $w(x)$ is supported by the cube $-\varepsilon \leq x_j \leq 1 + \varepsilon$, $1 \leq j \leq n$, for some positive ε, and if it satisfies

$$
\sum_{k \in \mathbf{Z}^n} |w(x-k)|^2 = 1 \qquad \text{identically} .
$$

We can assume that $w(x) \geq 0$.

Then the corresponding Gabor wavelets are the functions

$$
w_{k,l}(x) = (2\pi)^{-n/2}\, w(x-k)e^{il \cdot x} , \quad l \in \mathbf{Z}^n , \; k \in \mathbf{Z}^n .
$$

An elementary application of Fourier series yields

$$
(7.1) \qquad\qquad f(x) = \sum_k \sum_l \langle f, w_{k,l} \rangle w_{k,l}(x)
$$

and one easily checks that $w_{k,l}$ is a frame.

From now on it will be assumed that w belongs to $C_0^\infty(\mathbb{R}^n)$. Then the Fourier transform of $w_{k,l}$ is given by

$$
(7.2) \qquad\qquad \hat{w}_{k,l}(\xi) = (2\pi)^{-n/2}\, \hat{w}(\xi - l)\, e^{-ik\xi} .
$$

In other terms the Fourier transforms of our Gabor wavelets do have a similar structure. However $\hat{w}(\xi)$ is no longer compactly supported but has rapid decay at infinity together with all its derivatives.

Let us compute the entries $\langle Tw_{k,l}, w_{k',l'}\rangle$ when T is a pseudo-differential operator with a symbol $\sigma(x,\xi)$ fulfilling (4.3) and when $\delta = 0$. We have

$$(7.3) \qquad Tf(x) = (2\pi)^{-n} \int_{\mathbb{R}^n} e^{ix\cdot\xi}\, \sigma(x,\xi)\, \hat{f}(\xi)\, d\xi$$

and

$$\langle Tf, g\rangle = (2\pi)^{-n} \iint e^{ix\cdot\xi}\, \sigma(x,\xi)\, \hat{f}(\xi)\, \overline{g}(x)\, d\xi\, dx\ .$$

When $f = w_{k,l}$ and $g = w_{k',l'}$, we obtain

$$\langle Tf, g\rangle = \gamma\, I(k,l,k',l') \qquad \text{where} \quad |\gamma| = (2\pi)^{-2n}$$

and

$$I(k,l,k',l') = \iint e^{i(k'-k)\xi}\, e^{-i(l'-l)x}\, \theta_{(k,l,k',l')}(x,\xi)\, dx\, d\xi$$

where

$$\theta_{(k,l,k',l')}(x,\xi) = \hat{w}(\xi)\, w(x)\, e^{ix\xi}\, \sigma(x+k'\, ,\, \xi+l)\ .$$

Then $\theta_{(k,l,k',l')}$ belongs to a bounded subset of $\mathcal{S}(\mathbb{R}^n \times \mathbb{R}^n)$ which implies

$$|I(k,l,k',l')| \leq C\big(|k'-k| + |l'-l|\big)^{-2n-1}\ .$$

Therefore Schur's lemma can be applied.

Next we consider the same problem treated with Cotlar's lemma. The beginning of the story is almost the same. One starts with a partitioning of the "time frequency plane" or "phase space" given by $1 = \sum_k \sum_l \varphi(x-k)\,\varphi(x-l)$ where $\varphi \in C_0^\infty(\mathbb{R}^n)$. Then one decomposes the symbol $\sigma(x,\xi)$ into the building blocks $\varphi(x-k)\,\varphi(\xi-l)\,\sigma(x,\xi) = \sigma_{(k,l)}(x,\xi)$ and consider the corresponding operators $T_{k,l}$. Finally one applies Cotlar's lemma to the decomposition $T = \sum_k \sum_l T_{k,l}$.

We need to compute the operator norms of $T_{k',l'}^* T_{k,l}$ and $T_{k',l'} T_{k,l}^*$. To perform this task, we first compute the associated kernels. We have

$$T_{k,l}f(x) = \int K_{(k,l)}(x,y)\, f(y)\, dy$$

where

$$\begin{aligned} K_{(k,l)}(x,y) &= (2\pi)^{-n} \int e^{i\xi(x-y)}\, \sigma_{(k,l)}(x,\xi)\, d\xi \\ &= e^{il(x-y)}\, Z_{(k,l)}(x+k\, ,\, y+k)\ . \end{aligned}$$

These functions $Z_{(k,l)}(x,y)$ belong to a bounded set of $\mathcal{S}(\mathbb{R}^n \times \mathbb{R}^n)$ which implies that the kernels $K_{(k,l)}(x,y)$ have a sharp phase space localization.

Finally the kernel of $T_{k',l'} T_{k,l}^*$ is

$$\begin{aligned} A_{(k,l,k',l')}(x,y) &= \int K_{(k',l')}(x,u)\, \overline{K}_{(k,l)}(y,u)\, du \\ &= c\, e^{i(l'x-ly)} \int e^{i(l-l')u}\, Z_{(k',l')}(x+k',u+k')\, \overline{Z}_{(k,l)}(y+k,u+k)\, du\ . \end{aligned}$$

We have $|c| = 1$ and this kernel has fast decay when $|l' - l| + |k' - k|$ tends to infinity. To apply Cotlar's lemma, one needs to bound

$$\sup_x \int \left| A_{(k,l,k',l')}(x,y) \right| dy \qquad \text{and} \qquad \sup_y \int \left| A_{(k,l,k',l')}(x,y) \right| dx .$$

These two quantities still have rapid decay as $|l' - l| + |k' - k|$ tends to infinity.

In order to compare Cotlar's lemma to Schur's lemma, let us assume that instead of belonging to $C^\infty(\mathbb{R}^n \times \mathbb{R}^n)$ our symbol $\sigma(x,\xi)$ belongs to the Hölder class $C^r(\mathbb{R}^n \times \mathbb{R}^b)$ and ask ourselves the following question: what is the minimal amount of smoothness that permits to use Schur's lemma or Cotlar's lemma?

For Schur's lemma, the answer is $r > n$. We need to know if the Fourier coefficients of a function of $2n$ variables belong to l^1 and the answer to this problem is well-known (Beurling-Pollard's theorem).

When applying Cotlar's lemma, the computation looks similar but at the end we compute a sum of square roots (instead of a sum). Then the amount of smoothness which is needed is given by $r > 3n$. This is exactly what Calderón and Vaillancourt were assuming.

Note that in this example both approaches are superseded by a better method (which provides the optimal regularity) given in [**13**].

We continue to study pseudo-differential operators with Schur's lemma and consider the symbol class $S^0_{\delta,\delta}$ defined by (4.3) where $0 < \delta < 1$. For the sake of simplicity, the following discussion is limited to the one dimensional case. We first construct large families of frames providing discrete versions for a windowed Fourier analysis. Then the relation between windowed Fourier analysis and operators in $Op \, S^0_{\delta,\delta}$ will become clear.

Let $\omega_j, \, j \in \mathbf{Z}$, be an increasing sequence of real numbers such that $\lim_{j \to +\infty} \omega_j = +\infty$, $\lim_{j \to +\infty} \omega_j = -\infty$. We denote by η_j some positive real numbers such that $l_j = \omega_{j+1} - \omega_j \geq \eta_{j+1} + \eta_j$.

We first define some smooth windows $w_j(t)$. They satisfy

$$(7.4) \qquad\qquad \sum_{-\infty}^{\infty} |w_j|^2 = 1$$

$$(7.5) \qquad\qquad w_j(t) = 0 \quad \text{if} \quad t \leq \omega_j - \eta_j \quad \text{or} \quad t \geq \omega_{j+1} + \eta_{j+1}$$

$$(7.6) \qquad\qquad w_j(t) = 1 \quad \text{if} \quad \omega_j + \eta_j \leq t \leq \omega_{j+1} - \eta_{j+1}$$

$(7.7) \qquad\qquad w_j(t)$ is as smooth as possible, taking in account
the preceding constraints .

Next $L_j = l_j + \eta_j + \eta_{j+1}$ and we define wavelets $w_{j,k}(t)$ by

$$(7.8) \qquad\qquad w_{j,k}(t) = \frac{1}{\sqrt{L_j}} \, w_j(t) \, \exp\left(\frac{2k\pi it}{L_j}\right) \qquad j,k \in \mathbf{Z} .$$

This full collection is a tight frame [14] which means that, for any f in $L^2(\mathbb{R})$, one has

$$(7.9) \qquad f(t) \;=\; \sum_j \sum_k \langle f, w_{j,k} \rangle w_{j,k}(t) \;.$$

We now consider, in one dimension, an operator with symbol $\sigma(x, \xi)$ belonging to the symbol class $S^0_{1/2,1/2}$. We introduce a partitioning of the frequency axis into intervals $[\omega_j, \omega_{j+1}]$ defined by $\omega_j = j^2$ if $j = 0, 1, 2, \ldots$ and $\omega_j = -j^2$ if $j = -1, -2, -3, \ldots$ We then define η_j by $\frac{j}{10}$ when $j \geq 1$, $\eta_0 = \frac{1}{10}$ and $\eta_j = \frac{|j|}{10}$ if $j \leq -1$.

The windows $w_j(t)$ are defined as in (7.4)-(7.7) and the optimal smoothness in our present situation is $\left| \left(\frac{d}{dt} \right)^m w_j(t) \right| \leq C_m |j|^{-m}$ which is obviously compatible with (7.4)–(7.7).

Finally the wavelets $\psi_{j,k}$, $j \in \mathbf{Z}$, $k \in \mathbf{Z}$, we want to consider are defined by

$$(7.10) \qquad \psi_{j,k}(\xi) \;=\; \frac{1}{\sqrt{L_j}} \, w_j(\xi) \exp\left(\frac{2\pi i k \xi}{L_j} \right)$$

where $j \in \mathbf{Z}$, $k \in \mathbf{Z}$ and $L_j = \frac{11}{10}(2j+1)$ if $j \geq 0$, $L_{-j} = L_{j-1}$ if $j \geq 1$.

It is easily checked that operators with symbols in the class $S^0_{1/2,1/2}$ are almost diagonal in this basis. More precisely one has

$$\sup_{(j',k')} \sum_{(j,k)} |\langle T\psi_{j,k}, \psi_{j',k'} \rangle| \;\leq\; C_0$$

and

$$\sup_{(j,k)} \sum_{(j',k')} |\langle T\psi_{j,k}, \psi_{j',k'} \rangle| \;\leq\; C_1 \;.$$

The property which is relevant for proving these estimates is the fact that T "behaves nicely" on functions f which admit the following localization in the "time-frequency plane": the "support" of f is the interval $\left[\frac{k}{|j|}, \frac{k+1}{|j|} \right]$ and the "support of the Fourier transform" of f is the interval $[\omega_j, \omega_{j+1}]$. This localization is impossible the way it is stated; we mean in fact that if f is rescaled into $f(|j|t - k)\, e^{-ijt} = g_{j,k}(t)$, then the functions $g_{j,k}$ should belong to a bounded subset of $\mathcal{S}(\mathbb{R})$. If $g_{j,k}$ belongs to such a bounded set $\mathcal{B}$ of the Schwartz class, then we write $f \in \mathcal{B}_{j,k}$; this property is preserved by the action of the operator T uniformly in j and k, and this then leads to the estimates above.

From this discussion it is easy to guess what the "time frequency atoms" should look like in the case of operators with symbols in $S^0_{\delta,\delta}$. The partitioning of the frequency axis should be given by $\omega_j = \pm |j|^\alpha$ where $\alpha = \frac{1}{1-\delta}$, $\pm$ is the signum of j and the corresponding wavelets are constructed as in the special case $\delta = 1/2$.

What these examples show is that applying Schur's lemma needs some expertise and some serious knowledge about the operators to be studied.

The last example we would like to consider is due to S. Semmes [25] and provides a second solution to the problem raised by A. Zygmund. We want to prove that the Hilbert transform is bounded on $L^2(\mathbb{R})$ without using the Fourier transformation. The simplest way is to apply Schur's lemma to the Haar basis with a weight which will be defined.

The Haar basis was constructed in 1909 by A. Haar. It is an orthonormal basis for $L^2(\mathbb{R})$ with an amazingly simple structure. We start with the function $h(x)$ defined by $h(x) = 1$ on $[0, 1/2)$, $h(x) = -1$ on $[1/2, 1)$ and $h(x) = 0$ elsewhere. Next we consider the collection $\mathcal{I}$ of all dyadic intervals $I = [k2^{-j}, (k+1)2^{-j})$, $k \in \mathbf{Z}$, $j \in \mathbf{Z}$, and for each such interval, we define $h_I(x)$ by $2^{j/2} h(2^j x - k)$. Then the support of h_I is exactly I, $h_I(x) = |I|^{-1/2}$ on the left half of I and $h_I(x) = -|I|^{-1/2}$ on the right half of I.

This collection $h_I(x)$, $I \in \mathcal{I}$, is an orthonormal basis for $L^2(\mathbb{R})$ which has some other remarkable properties. For $1 < p < \infty$ the Haar basis is an unconditional basis for $L^p(\mathbb{R})$. It means that the expansion of a function $f \in L^p(\mathbb{R})$ into the Haar basis is a summable family. This is fortunate since there is no natural ordering for the set $\mathcal{I}$.

There is a second version of the Haar basis (which is the one considered by A. Haar). This version is an orthonormal basis for $L^2[0, 1)$, and an unconditional basis for $L^p[0, 1)$ for $1 < p < \infty$. (Note that the trigonometric system which also gives an orthonormal basis for $L^2[0, 1)$ is not an unconditional basis for $L^p[0, 1)$ if $p \neq 2$.) The Haar basis on $[0, 1)$ consists of the function 1 and of all the Haar functions $h_I(x)$ for which $I \subset [0, 1)$. Then there is a natural (lexicographical) ordering which amounts to writing $n = 2^j + k$, $j \geq 0$, $0 \leq k < 2^j$, and defining $h_n(x) = 2^{j/2} h(2^j x - k)$; if $n = 0$, $h_0(x) = 1$. A. Haar proved in his thesis that for any continuous functions f on $[0, 1]$, the expansion of f in the Haar system which reads $\langle f, h_0 \rangle h_0(x) + \cdots + \langle f, h_n \rangle h_N(x) + \cdots$ is uniformly convergent to $f(x)$. This also is a property that the trigonometric system does not possess.

S. Semmes proved that Schur's lemma works for the Hilbert transform with the Haar basis and the weight $|I|^{1/2}$.

8. Orthonormal wavelet bases.

Let us take the example of the expansion in the Haar basis of a function belonging to $L^2(\mathbb{R})$. If this function $f(x)$ happens to belong to the Sobolev space $H^1(\mathbb{R})$ we could be tempted to differentiate this expansion term by term. We then obtain an inappropriate expansion of an L^2 function $f'(x)$ into a series of Dirac masses.

Wavelet bases have the same structure as the Haar system but offer the amazing improvement that the corresponding expansion can be differentiated term by term if the function $f(x)$ is smooth. This is also an important property of the trigonometric system.

In some sense wavelets behave like eigenfunctions of the differential operator

$\frac{d}{dx}$; this statement was given a precise formulation by C. Albanese [1].

Let us return to specific definitions. There are two versions of the Haar system for $L^2(\mathbb{R})$. The first one is the full collection h_I, $I \in \mathcal{I}$. The second one is the string $\chi(x-k)$, $k \in \mathbf{Z}$, together with all $h_I(x)$ where $|I| \leq 1$; we have denoted by $\chi(x)$ the indicator function of the interval $[0,1)$. In this second version, a function $f(x)$ in $L^2(\mathbb{R})$ is written as a sketchy approximation $f_0(x) = \sum c_k \chi(x-k)$ plus layers of finer and finer details which are needed for describing the full complexity of $f(x)$.

These two versions of the Haar system are not fully equivalent. For instance if $f(x)$ is a smooth function with compact support, its expansion in the first Haar system will not converge to $f(x)$ with respect to the L^1-norm. Otherwise we could integrate term by term and would obtain that the integral of $f(x)$ is always 0. However in the second version, the expansion converges to $f(x)$ with respect to the L^1-norm whenever f belongs to $L^1(\mathbb{R})$.

Similarly there are two possibilities for defining orthonormal wavelet bases.

In the first case, the function $h(x)$ which is used in the Haar system is replaced by a carefully designed function $\psi(x)$ enjoying the following properties

$$(8.1) \qquad \left| \left(\frac{d}{dt} \right)^q \psi(x) \right| \leq C_m (1 + |x|)^{-m} \qquad \text{for} \quad 0 \leq q \leq r \,, \; m \geq 1$$

(in the Haar basis case, $r = 0$)

$$(8.2) \qquad \psi_{j,k}(x) = 2^{j/2} \psi(2^j x - k) \quad , \quad j \in \mathbf{Z} \,, \; k \in \mathbf{Z} \,,$$
$$\text{is an orthonormal basis for } L^2(\mathbb{R}) \,.$$

It can be easily proved that these two properties imply a third one

$$(8.3) \qquad \int x^q \, \psi(x) \, dx = 0 \qquad \text{for} \quad 0 \leq q \leq r \,.$$

If I denotes the dyadic interval $[k2^{-j}, (k+1)2^{-j})$, $\psi_I = \psi_{j,k}$ is no longer supported by I but has fast decay outside I, at the scale given by $|I|$.

The first orthonormal wavelet basis was constructed by J.O. Strömberg in 1981. This construction based on special properties of spline functions could not be generalized to other settings and it is at the same time unfortunate and fortunate that Strömberg's paper was not given the credit it deserved.

The second option for defining orthonormal wavelet bases is to start with two functions $\varphi(x)$ and $\psi(x)$. The function $\varphi(x)$ is named the "scaling function". It satisfies (8.1) but not (8.3). Instead we have

$$(8.4) \qquad x^q = \sum c(q,k) \, \varphi(x-k)$$

where $c(q,k) = \int x^q \, \overline{\varphi(x-k)} \, dx$ and q is integer, $0 \leq q \leq r$. Moreover, we now replace (8.2) by the requirement that $\varphi(x-k)$, $k \in \mathbf{Z}$, together with $2^{j/2} \psi(2^j x - k)$, $j \geq 0$, $k \in \mathbf{Z}$, be an orthonormal basis for $L^2(\mathbb{R})$.

In the second option, one has to construct φ and ψ. But ψ can be easily deduced from φ and this relation is clarified by a third approach to orthonormal wavelet bases.

A multiresolution analysis is defined as an increasing sequence V_j, $j \in \mathbf{Z}$, of closed subspaces of $L^2(\mathbb{R})$ with the four following properties

(8.5) $\displaystyle\bigcup_{-\infty}^{\infty} V_j \quad \text{is dense in} \quad L^2(\mathbb{R}) \quad \text{and} \quad \bigcap_{-\infty}^{\infty} V_j = \{0\}$

(8.6)   whenever $f(x)$ belongs to V_j, $f(2x)$ belongs to V_{j+1}

(8.7)   whenever $f(x)$ belongs to V_0, so does $f(x-k)$ for every $k \in \mathbf{Z}$

(8.8)   there exists a function $g(x)$ in V_0 such that
             $g(x-k)$, $k \in \mathbf{Z}$, is a Riesz basis for V_0 .

A Riesz basis for a Hilbert space H is an unconditional basis (see definition 6.1). It would not be more restrictive to replace "Riesz basis" in (8.8) by "orthonormal basis": the Riesz basis $g(x-k)$, $k \in \mathbf{Z}$, is easily transformed into an orthonormal basis $\varphi(x-k) = \sum_l \omega(k-l)\, g(x-l)$ for V_0.

A multiresolution analysis is r-regular if these exists a choice for g such that $\left|\left(\frac{d}{dt}\right)^q g(x)\right| \le C_m(1+|x|)^{-m}$ for $0 \le q \le r$ and every $m \ge 1$. Then the coefficients $\omega(k)$, $k \in \mathbf{Z}$, can be fixed in such a way that φ still fulfills these smoothness and decay properties.

The embedding $V_0 \subset V_1$ immediately yields

(8.9) $\varphi(x) \;=\; 2\sum \alpha_k\, \varphi(2x - k)$

where the α_k decay rapidly for k tending to infinity. The corresponding wavelet $\psi(x)$ is defined by

(8.10) $\psi(x) \;=\; 2\sum \beta_k\, \varphi(2x - k) \qquad \text{where} \quad \beta_k = (-1)^k \overline{\alpha}_{1-k}$.

Then $\psi(x-k)$, $k \in \mathbf{Z}$, together with $\varphi(x-k)$, $k \in \mathbf{Z}$, constitutes an orthonormal basis for V_1. In other terms $\psi(x-k)$, $k \in \mathbf{Z}$, is an orthonormal basis for the orthogonal complement W_0 of V_0 inside V_1. Let us denote by W_j the orthogonal complement of V_j inside V_{j+1}. By (8.5) we have

(8.11) $\displaystyle L^2(\mathbb{R}) \;=\; \bigoplus_{-\infty}^{\infty} W_j$.

By a simple rescaling, we obtain that $2^{j/2}\,\psi(2^j x - k)$, $k \in \mathbf{Z}$, is an orthonormal basis for W_j. This last observation, together with (8.11), implies also that $\psi_{j,k}$, $j \in \mathbf{Z}$, $k \in \mathbf{Z}$, is an orthonormal basis for $L^2(\mathbb{R})$.

A multiresolution analysis is a concept which is appealing to approximation theorists. To give an example of a multiresolution analysis, let us denote by V_0 the subspace of $L^2(\mathbb{R})$ consisting of cubic splines with knots $k \in \mathbf{Z}$. Then the corresponding scaling function $\varphi(x)$ was well known to spline people since they were interested in the properties of the orthogonal projector $P_j : L^2(\mathbb{R}) \to V_j$. However J.O. Strömberg was the first mathematician who introduced the corresponding wavelet $\psi(x)$ in this spline context.

A fourth approach to orthonormal wavelet bases is given by algorithms which have been created by signal processing experts and are known as quadrature mirror filters or subband coding.

The equivalence between these four approaches is a program which was launched by S. Mallat, further developed by I. Daubechies, and successfully completed by A. Cohen and P. G. Lemarié-Rieusset. A. Cohen studied the equivalence between the multiresolution analysis approach and subband coding [3]. P.G. Lemarié-Rieusset proved that for any orthonormal basis of compactly supported wavelets fulfilling (8.1) there always exists a scaling function $\varphi(x)$; the restriction of compact support was later lifted by P. Auscher [7] and P.G. Lemarié-Rieusset [19].

Using this equivalence, I. Daubechies proved that, given any $r \geq 0$, there exists a compactly supported scaling function φ, of class C^r, as well as a compactly supported wavelet ψ of class C^r. Taking the Hilbert transform as an example again, using these Daubechies wavelets instead of the Haar system leads to improved off-diagonal decay of the entries $\langle H\psi_I, \psi_J \rangle$.

In $\mathbb{R}^n$ there are two approaches to orthonormal wavelet bases. The first one is a straightforward extension of the one dimensional case. If $\varphi(t)$ and $\psi(t)$ are defined as above, then their n-dimensional analogues are given by $\varphi(x) = \varphi(x_1)\varphi(x_2)\ldots\varphi(x_n)$ when $x = (x_1,\ldots,x_n)$ and $\psi_\varepsilon(x) = \psi_{\varepsilon_1}(x_1)\ldots\psi_{\varepsilon_n}(x_n)$ where $\varepsilon = (\varepsilon_1,\ldots,\varepsilon_n) \in \{0,1\}^n$, $\varepsilon \neq (0,0,\ldots,0)$ and $\psi_0(t) = \varphi(t)$, $\psi_1(t) = \psi(t)$ by convention. It means that there are $2^n - 1$ distinct wavelets in the n-dimensional setting. This construction gives an important role to the coordinates axes. There are some other options. One is more rotationally invariant and is useful in the study of the n-dimensional fractional Brownian motion. In this option, $V_0 \subset L^2(\mathbb{R}^2)$ is defined by $f \in L^2(\mathbb{R}^2)$ and $\Delta^3 f = \sum_{k \in \mathbf{Z}^2} c_k \delta_k$ where δ_k is the Dirac mass at k and $\sum |c_k|^2 < \infty$. The V_j are deduced from V_0 by the same rule as in (8.6). We denote by W_j the orthogonal complement of V_j inside V_{j+1} and there exist three wavelets ψ_1, ψ_2 and ψ_3 such that $\{\psi_1(x - k), \psi_2(x - k), \psi_3(x - k) \; ; \; k \in \mathbf{Z}^2\}$ constitutes an orthonormal basis for W_0. It follows that $\{2^j \psi_q(2^j x - k) \; ; \; 1 \leq q \leq 3 \;, \; j \in \mathbf{Z} \;, \; k \in \mathbf{Z}^2\}$ is an orthonormal basis for $L^2(\mathbb{R}^2)$.

We now return to operator theory.

9. Generalized Calderón-Zygmund operators.

A. Calderón studied a remarkable class of singular integral operators. This class is defined by $Tf(x) = \text{p.v.} \int K(x,y) f(y) \, dy$ where $K(x,y)$ satisfies the three following properties:

$$(9.1) \qquad\qquad |K(x,y)| \leq C_0 |x - y|^{-n}$$

(9.2) $$K(y, x) = -K(x, y)$$

(9.3)   there exists an exponent $\gamma \in (0, 1)$ and a constant
             C such that for $\ |x' - x| \leq \frac{1}{2}|x - y|$
             $$|K(x', y) - K(x, y)| \leq C|x' - x|^{\gamma}|x - y|^{-n-\gamma}.$$

Our goal will be to show that wavelet based methods permit a complete description of this class of operators. These operators are not always bounded on $L^2(\mathbb{R}^n)$ but the wavelet analysis which will be applied leads to a necessary and sufficient condition for boundedness.

The first temptation is to apply Schur's lemma. If f and g are two testing functions, we have

$$
\begin{aligned}
\langle Tf, g \rangle &= \text{p.v.} \iint K(x, y)\, f(y)\, g(x)\, dx\, dy \\
&= \lim_{\varepsilon \downarrow 0} \iint_{|y-x| \geq \varepsilon} K(x, y)\, f(y)\, g(x)\, dy\, dx \\
&= \frac{1}{2} \iint K(x, y)\, [f(y)g(x) - f(x)g(y)]\, dy\, dx
\end{aligned}
$$

and this last integral is absolutely convergent.

Another important observation is the fact that whenever $K(x, y)$ satisfies (9.1), (9.2) and (9.3), so does $\alpha^n K(\alpha x + x_0, \alpha y + x_0)$ for every $\alpha > 0$ and $x_0 \in \mathbb{R}^n$. What is crucial is that the constants C_0, C_1 and the exponent γ are not altered by such rescalings.

Therefore is it extremely natural to compare $\langle Tf, g \rangle$ with $\langle Tf_{\alpha, x_0}, f_{\alpha, x_0} \rangle$ where $f_{\alpha, x_0}(x) = \alpha^{n/2} f(\alpha(x - x_0))$, $g_{\alpha, x_0}(x) = \alpha^{n/2} g(\alpha(x - x_0))$.

If T happens to be bounded on $L^2(\mathbb{R}^n)$, then we have $|\langle Tf_{\alpha, x_0}, g_{\alpha, x_0} \rangle| \leq \|T\|\, \|f\|_2\, \|g\|_2$ which means that computing such entries $\langle Tf_{\alpha, x_0}, g_{\alpha, x_0} \rangle$ is a good test for studying this boundedness.

A final remark concerns entries of the form $\langle Tf_{\alpha, x_0}, g_{\alpha, x_1} \rangle$ where f and g are two fixed smooth and compactly supported testing functions. When $|x_1 - x_0| \leq \alpha$, such an entry can always be rewritten $\langle Tf_{\alpha, x_0}, h_{\alpha, x_0} \rangle$ where h is g shifted by $\alpha^{-1}(x_1 - x_0)$. For that reason, $h(x)$ shares with $g(x)$ the same smoothness and localization. When $|x_1 - x_0|$ is much larger than α, one writes

$$\langle Tf_{\alpha, x_0}, g_{\alpha, x_1} \rangle = \iint K(x, y)\, f_{\alpha, x_0}(y)\, g_{\alpha, x_1}(x)\, dy\, dx.$$

If the integral of f vanishes, this double integral is rewritten $\iint [K(x, y) - K(x, x_0)] f_{\alpha, x_0}(y)\, g_{\alpha, x_1}(x)\, dy\, dx$ and is bounded by $C\alpha^{\gamma}\|f\|_1\|g\|_1|x_1 - x_0|^{-n-\gamma}$. This estimate is an improvement on $C\|f\|_1\|g\|_1|x_1 - x_0|^{-n}$ which is what we get if $\int f(x)\, dx = \int g(x)\, dx = 1$. In both cases we use that f and g are both compactly supported.

If one applies Schur's lemma to such an operator, we cannot conclude. The reason of the failure is that the operators we are studying might be unbounded. We will present in the following section another attack to this problem.

10. The non-standard representation.

Let us consider the n-dimensional wavelets ψ_λ which were defined in section 8. The index λ belongs to $\Lambda = \mathbf{Z} \times \mathbf{Z}^n \times E$ where $E = \{0,1\}^n \setminus \{(0,0,\dots,0)\}$ and $\lambda = (j,k,\varepsilon)$.

Applying Schur's lemma amounts to computing $\langle T\psi_\lambda, \psi_{\lambda'} \rangle$. As noted in "Ondelettes et Opérateurs II", p. 268, before computing this entry, one should make a few observations. When $\lambda(j,k,\varepsilon)$ and $\lambda' = (j',k',\varepsilon')$, the case $j = j'$ is the one which has been treated in the preceding section. The two wavelets ψ_λ and $\psi_{\lambda'}$ are living on the same scale and one obtains $|\langle T\psi_\lambda, \psi_{\lambda'} \rangle| \leq C(1 + |k-k'|)^{-n-\lambda}$ as was explained in the last section.

Assuming, for the sake of simplicity, that $K(x,y)$ is real valued, as well as the wavelets ψ_λ, one has

$$\langle T\psi_\lambda, \psi_{\lambda'} \rangle \;=\; -\langle \psi_\lambda, T\psi_{\lambda'} \rangle \;=\; -\langle T\psi_{\lambda'}, \psi_\lambda \rangle \,.$$

Therefore, if $j' \neq j$, one can always assume that $j' \leq j - 1$. Then $\psi_{\lambda'}$ belongs to V_j and before computing $\langle T\psi_\lambda, \psi_{\lambda'} \rangle$ one can replace $T\psi_\lambda$ by the orthogonal projection of $T\psi_\lambda$ onto V_j. This orthonormal projection is named $g(x)$ and is computed through

$$(10.1) \qquad\qquad g(x) \;=\; \sum_k \langle T\psi_\lambda, \varphi_{j,k} \rangle \varphi_{j,k}(x)$$

where, as usual, $\varphi_{j,k}(x) = 2^{j/2}\,\varphi(2^j x - k)$.

In fact, the entries we are calculating define the non-standard representation of our operator T. Let us be slightly more systematic: if T is any operator defined on a dense subset $\mathcal{V} \subset L^2(\mathbb{R}^n)$ containing both the scaling functions $\varphi_{j,k}$ and the wavelets ψ_λ, then the non standard representation of T is given by the three matrices

(10.2)

$$\begin{cases} A \;=\; \big(\langle T\psi_\lambda, \varphi_{j,l} \rangle\big) & \text{where} \quad \lambda = (j,k,\varepsilon) \in \Lambda \,,\; l \in \mathbf{Z}^n \\[2pt] B \;=\; \big(\langle T\varphi_{j,k}, \psi_\lambda \rangle\big) & \text{where} \quad \lambda = (j,k,\varepsilon) \in \Lambda \,,\; l \in \mathbf{Z}^n \\[2pt] \text{and} \\[2pt] C \;=\; \big(\langle T\psi_\lambda, \psi_{\lambda'} \rangle\big) & \text{where} \quad \lambda = (j,k,\varepsilon) \in \Lambda \,,\; \lambda' = (j,l,\varepsilon') \in \Lambda \,. \end{cases}$$

What is crucial in this approach is that the scale is always the same in ψ_λ as in $\varphi_{j,k}$, $\varphi_{j,l}$ or $\psi_{\lambda'}$. As was explained above, it is trivial to compute all the entries which are needed for applying Schur's lemma from these non standard matrices.

In order to keep the notations as simple as possible, we return to the one-dimensional case which already exhibits all the difficulties of the problem. We write $\alpha(j,k,l) = \langle T\psi_{j,k}, \varphi_{j,l} \rangle$, $\beta(j,k,l) = \langle T\varphi_{j,k}, \psi_{j,l} \rangle$, $\gamma(j,k,l) = \langle T\psi_{j,k}, \psi_{j,l} \rangle$.

Then we have

THEOREM 10.1. The operators T with kernels fulfilling (9.1), (9.2) and (9.3) are characterized by the following properties

(10.3)
$$\begin{cases} |\alpha(j,k,l)| \le C(1+|k-l|)^{-1-\gamma} \\ |\gamma(j,k,l)| \le C(1+|k-l|)^{-1-\gamma} \qquad \text{and} \\ \beta(j,k,l) = -\alpha(j,l,k) \\ \gamma(j,k,l) = -\gamma(j,l,k) \,. \end{cases}$$

The kernel $K(x,y)$ is given by

(10.4)
$$K(x,y) = \sum_j \sum_k \sum_l \alpha(j,k,l)\, \varphi_{j,l}(x)\, \psi_{j,k}(y)$$
$$+ \sum_j \sum_k \sum_l \beta(j,k,l)\, \psi_{j,l}(x)\, \varphi_{j,k}(y)$$
$$+ \sum_j \sum_k \sum_l \gamma(j,k,l)\, \psi_{j,l}(x)\, \psi_{j,k}(y) \,.$$

One can give to this decomposition a different interpretation in which $K(x,y)$ is viewed as a distribution belonging to $\mathcal{S}'(\mathbb{R}^2)$ and one uses the two-dimensional wavelets to expand this distribution as a wavelet series; the result is exactly (10.4). This interpretation is due to G. Beylkin, R. Coifman and V. Rokhlin.

There is a third interpretation for this expansion. Let us denote by $P_j :$ $L^2(\mathbb{R}) \to V_j$ the orthogonal projection onto the subspace V_j. Similarly $Q_j = P_{j+1} - P_j$ is the orthonormal projection onto W_j. With these notations, we have

(10.5)
$$T = \lim_{j\to+\infty} P_j\, T\, P_j - \lim_{j\to-\infty} P_j\, T\, P_j \,.$$

At least this is true if T is bounded from $H^1(\mathbb{R})$ into $L^2(\mathbb{R})$. This approximation to T can be rewritten as a telescopic series

$$T = \sum_{-\infty}^{\infty} (P_{j+1}\, T\, P_{j+1} - P_j\, T\, P_j)$$

which leads to

(10.6)
$$T = \sum_{-\infty}^{\infty} P_j\, T\, Q_j + \sum_{-\infty}^{\infty} Q_j\, T\, P_j + \sum_{-\infty}^{\infty} Q_j\, T\, Q_j \,.$$

These series are exactly the three series that show up in the expansion of the kernel $K(x,y)$ of T.

The problem of studying the L^2-continuity of T will be postponed until section 12. A detour is needed and we will carefully study a special case of operators T given by (10.4).

11. Paraproducts.

Among operators T given by (10.3) or (10.4) there is a subclass which is amazingly simple and natural. We will ignore condition (9.2) and assume that $\alpha(j,k,l) = 0$ for $k \neq l$, $\beta(j,k,l) = \gamma(j,k,l) = 0$ for all j, k and l. Then the corresponding operator is defined by

$$(11.1) \qquad\qquad T(\psi_{j,k}) \; = \; \lambda_{j,k}\,\varphi_{j,k}$$

where this sequence $\lambda_{j,k}$ of scalars completely characterizes T. The same question can be asked for the more general description

$$(11.2) \qquad\qquad T(\psi_{j,k}) \; = \; f_{j,k}$$

where $f_{j,k}$ is a given sequence in $L^2(\mathbb{R})$. The problem of finding necessary and sufficient conditions on this sequence $f_{j,k}$ for the L^2-continuity of this more general T cannot be accessed by human beings.

However it is striking that in the special case where $f_{j,k} = \lambda_{j,k}\,\varphi_{j,k}$ necessary and sufficient conditions exist. We write $\lambda(I) = 2^{-j/2}\,\lambda_{j,k}$ if $I = [k2^{-j}, (k+1)2^{-j})$ belongs to the family $\mathcal{I}$ of all dyadic intervals.

We say that $\lambda(I)$, $I \in \mathcal{I}$, satisfies the Carleson condition if there exists a constant C such that for every dyadic interval J one has

$$(11.3) \qquad\qquad \sum_{I \subset J} |\lambda(I)|^2 \; \leq \; C|J| \,.$$

With these notations, we have

THEOREM 11.1. The operator T given by (11.1) is bounded on $L^2(\mathbb{R})$ if and only if $\lambda(I) = 2^{-j/2}\,\lambda(j,k)$ satisfies the Carleson condition (11.3).

The L^2-boundedness of such operators T cannot be obtained by Schur's lemma.

We would like to provide the reader with a second approach to the class of bounded operators defined by (11.1). A function $b(x)$ of the real variable x belongs to the John and Nirenberg space $BMO(\mathbb{R})$ if $b(x)$ belongs to $L^2_{\mathrm{loc}}(\mathbb{R})$ and if there exists a constant C such that for every interval I (with length $|I|$) one has

$$(11.4) \qquad\qquad \frac{1}{|I|} \int_I |b(x) - m_I(b)|^2 \, dx \; \leq \; C \,.$$

Here $m_I(b)$ denotes the mean value of b over I. Returning to our operator T, $\sum\sum \lambda_{j,k}\,\psi(2^j x - k)$ belongs to $BMO(\mathbb{R})$ if and only if (11.3) holds. Then

$$T(f) \; = \; \sum_I \langle f, \psi_I \rangle \langle b, \psi_I \rangle |I|^{-1/2}\,\varphi_I(x)$$

and its adjoint T^* is given by

$$T^*(f) = \sum_I \langle b, \psi_I \rangle \langle f, \varphi_I \rangle |I|^{-1/2} \, \psi_I(x) \, .$$

Such an operator T^* is called the paraproduct between the BMO function $b(x)$ and the L^2 function f. It should be observed that one cannot multiply a BMO function with an L^2 function and expect to still obtain an L^2 function. Indeed $\log |x|$ belongs to $BMO(\mathbb{R})$, $\frac{1}{|x|^{1/2} \log |x|}$ belongs to $L^2(\mathbb{R})$ but their product does not belong to $L^2(\mathbb{R})$. However, the paraproduct does for you what the ordinary product cannot.

12. Reduction to paraproducts.

Let us return to these singular integral operators that Calderón wanted to study. For the sake of simplicity, $K(x, y)$ will be assumed to be real valued as well as φ and ψ. The problem we want to solve is to know whether or not T is bounded.

This problem has a straightforward answer in terms of the entries

$$\alpha(j, k, l) = \langle T\psi_{j,k}, \varphi_{j,l} \rangle \, .$$

We form $\tilde{\alpha}(j, k) = \sum_l \alpha(j, k, l)$. This series is convergent by (10.3). We now consider the paraproduct $\tilde{T}$ defined by $\tilde{T}(\psi_{j,k}) = \tilde{\alpha}(j, k) \, \varphi_{j,k}$ and write

$$(12.1) \qquad\qquad T = \tilde{T} - \tilde{T}^* + R \, .$$

We then have

THEOREM 12.1. The operator R defined by (12.1) is always bounded on $L^2(\mathbb{R})$ and the necessary and sufficient condition for the L^2-boundedness of T is that $\lambda(I) = 2^{-j/2} \, \tilde{\alpha}(j, k)$ should fulfill the Carleson condition (11.3).

Schur's lemma applies to the proof of the L^2-boundedness of R.

13. Compression for Calderón-Zygmund operators.

The result we would like to describe now was obtained by Yang Xi Qiang in his Ph.D. dissertation.

Let us return to the entries $\alpha(j, k, l)$, $\beta(j, k, l)$ and $\gamma(j, k, l)$ of the matrices A, B and C defined by (10.2). These entries have a fast off diagonal decay which is given by (10.3) and is uniform in j. In the preceding sections γ was assumed to belong to $(0, 1)$. But instead γ can be any positive number. One writes $\gamma = q + r$, $q \geq 1$, $0 < r < 1$ and (9.3) is then replaced by

$$(13.1) \qquad |\partial^\alpha K(x', y) - \partial^\alpha K(x, y)| \leq C|x' - x|^r \, |x - y|^{-n-\gamma}$$

when $|\alpha| = q$ and $|x' - x| \leq \frac{1}{2} |x - y|$.

Then Theorem 10.1 can easily be extended and (10.3) characterizes for any $\gamma > 0$, $\gamma \notin \mathbb{N}$, the entries of the non standard representation. A very natural approximation to A, B and C is given by the following procedure. For any $m \geq 1$, one replaces by 0 all the entries $\alpha(j,k,l)$, $\beta(j,k,l)$ and $\gamma(j,k,l)$ such that $|k - l| \geq m + 1$.

But this simple-minded procedure completely destroys the cancellations which provide the boundedness of T. A simple example where this occurs is the following. We define $\alpha(j,k,l)$ by $\alpha(j,k,k) = 2$ and $\alpha(j,k,l) = -2^{-|k-l|}$ if $k \neq l$. Then $\sum_l \alpha(j,k,l) = 0$. If $\alpha_m(j,k,l) = \alpha(j,k,l)$ when $|k - l| \leq m$ and $\alpha_m(j,k,l) = 0$ if $|k - l| > m$, we then obtain $\sum_l \alpha(j,k,l) = 2^{-m+1}$ and A_m will correspond to an unbounded operator.

This explains why $\alpha_m(j,k,l)$ and the corresponding matrix will be defined in a slightly different way:

(13.2)
$$\begin{cases} \alpha_m(j,k,l) = 0 & \text{if} \quad |k-l| > m \\ \alpha_m(j,k,l) = \alpha(j,k,l) & \text{if} \quad 1 \leq |k-l| \leq m \\ \text{and} \\ \alpha_m(j,k,k) = \alpha(j,k,k) + \displaystyle\sum_{|k-l|>m} \alpha(j,k,l) . \end{cases}$$

Then

$$\sum_l \alpha_m(j,k,l) = \sum_l \alpha(j,k,l) .$$

The same procedure is applied to $\beta(j,k,l)$; $\gamma(j,k,l)$ is treated straightforwardly.

The three banded matrices A_m, B_m and C_m provide the approximation T_m to T we were looking for. The quality of this approximation is given by the following theorem.

THEOREM 13.1. There is a constant $C = C(n,\gamma)$ such that for any operator T with a kernel satisfying (9.1), (9.2) and (9.3) one has, for each $m \geq 1$,

(13.3)
$$\|T - T_m\| \leq C\, m^{-\gamma} \sqrt{\log m} .$$

Moreover this estimate is optimal: there exists a bounded operator $T : L^2(\mathbb{R}^n) \to L^2(\mathbb{R}^n)$ with a kernel satisfying (9.1), (9.2) and (9.3) and a positive constant c such that for each $m \geq 1$

(13.4)
$$\|T - T_m\| \geq c\, m^{-\gamma} \sqrt{\log m} .$$

The first part of theorem remains true even if T is unbounded. An example of T satisfying (13.4) is given by $\alpha(j,k,l) = -|k-l|^{-1-\gamma}$ when $k \neq l$, $\sum_l \alpha(j,k,l) = 0$, $\beta(j,k,l) = -\alpha(j,k,l)$ and finally $\gamma(j,k,l) = 0$.

14. Conclusion.

Our program was to analyze large classes of operators through their representations in either "time-frequency" frames or in adapted orthonormal bases. Then one hopes to obtain almost diagonal matrices. An almost diagonal matrix can be studied by Schur's lemma.

We have also seen that this program has its limitations. For instance the multiplication by a function $m(x) \in L^\infty(\mathbb{R}^n)$ is not an almost diagonal operator in a wavelet basis. This is the reason why we gave up the standard representation and why the non standard representation is better suited to general Calderón-Zygmund theory.

REFERENCES

1. C. Albanese. Private communication.
2. B.K. Alpert. *Construction of simple multiscale bases for fast matrix operations.* In Wavelets and their applications, ed. by Mary Beth Ruskai, Jones and Bartlett, (1992).
3. P. Auscher. *Solution of two problems on wavelets.* (Preprint, IRMAR, Univ. Rennes I, 35042 Rennes Cedex).
4. G. Battle. *Wavelet refinement of the Wilson recursion formula.* Preprint, Math. Dept. Texas A & M University, College Station, TX 77843.
5. G. Beylkin, R. Coifman and V. Rokhlin. *Fast wavelet transforms and numerical algorithms, I.* Comm. Pure Appl. Math. Vol. XLIV (1991), 141–183.
6. J.M. Bony. *Propagation et interaction des singularités pour les solutions des équations aux dérivées partielles non-linéaires.* Proc. I.C.M. (1983), Warszawa, 1143–1147.
7. A. Cohen. *Ondelettes et traitement numérique du signal.* RMA (Recherches en Mathématiques Appliquées) 25. Masson (Paris) (1992).
8. A. Cohen and J.P. Conze. *Régularité des bases d'ondelettes et mesures ergodiques.* Revista Matemática Iberoamericana Vol. 8, n. 3, (1992), 351–365.
9. R. Coifman. *Adapted multiresolution analysis, computation, signal processing and operator theory.* Proceedings of the ICM Kyoto (1990), 879–887.
10. R. Coifman, P. Jones and S. Semmes. *Two elementary proofs of the L^2-boundedness of the Cauchy integral on Lipschitz curves.* Journal of the AMS **2** (1989), 553–564.
11. R. Coifman and R. Rochberg. *Representation theorems for holomorphic and harmonic functions in L^p.* Astérisque **77**, Sociéte Mathématique Française (1980).
12. R. Coifman and G. Weiss. *Analyse harmonique non-commutative sur certains espaces homogènes.* Lecture Notes in Mathematics, **242** (1971), Springer-Verlag.
13. R. Coifman and Y. Meyer. *Au-delà des opérateurs pseudo-différentiels.* Astérisque **57**, Sociéte Mathématique Française (1978).
14. I. Daubechies. *Ten lectures on wavelets.* CBMS-NSF. Regional conference series in applied mathematics, SIAM (1992).
15. G. David. *Wavelets and singular integrals on curves and surfaces.* Lecture Notes in Mathematics 1465, Springer-Verlag.
16. G. Fix and G. Strang. *Fourier analysis of the finite element method in Ritz-Galerkin theory.* Stud. Appl. Math. **48** (1969), 265–273.
17. L. Greengard and V. Rokhlin. *Rapid evaluation of potential fields in three dimensions.* Vortex Methods Proceedings (Los Angeles, 1987). Lecture Notes in Mathematics.
18. A. Knapp and E. Stein. *Intertwining operators on semi-simple Lie groups.* Ann. Math. **93**, (1971), 489–578.
19. P.G. Lemarié-Rieusset. *Sur l'existence des analyses multirésolutions en théorie des ondelettes.* Revista Matemática Iberoamericana, Vol. 8, n. 3, (1992), 457–474.
20. D. Marr. *Vision.* W.H. Freeman (1982).
21. Y. Meyer. *Ondelettes et opérateurs, tomes I, II & III.* Hermann, Paris, (1990).

22. Y. Meyer. *Wavelets and operators.* Cambridge studies in advanced mathematics, **37**, (1993).
23. Y. Meyer. *Les ondelettes, algorithmes et applications.* Armand Colin (1992).
24. Y. Meyer. *Ondelettes et algorithmes concurrents.* Hermann (1993).
25. S. Semmes. *Nonlinear Fourier analysis.* Bull. AMS, **20**, 1, Jan. 1989, 1–18.

CEREMADE, Université Paris-Dauphine, 75775 Paris Cédex 16, France
E-mail: ymeyer@dmi.ens.fr

Proceedings of Symposia in Applied Mathematics
Volume **47**, 1993

Projection Operators in Multiresolution Analysis

PIERRE GILLES LEMARIE-RIEUSSET

ABSTRACT. We describe various ways to deal with a bi-orthogonal multiresolution analysis. Different but equivalent points of view focus on one of the following: the scaling functions, the scaling filters, the wavelets or the large-scales projection operator P_0. This projector itself can be described in many ways, as a multi-scale operator dealing with wavelets, as an one-scale projection operator dealing with scaling functions or as a shift-invariant projector dealing with an over-sampled family of wavelets.

1. Bi-orthogonal multiresolution analysis.

We first recall some definitions and notations that we will use throughout the whole paper.

A *multiresolution analysis* (as defined by S. Mallat [**22**] in 1986) is a sequence $(V_j)_{j \in \mathbb{Z}}$ of closed linear subspaces of $L^2(\mathbb{R})$ such that:

(1.1) $V_j \subset V_{j+1}$, $\cap_{j \in \mathbb{Z}} V_j = \{0\}$ and $\cup_{j \in \mathbb{Z}} V_j$ is dense in $L^2(\mathbb{R})$

(1.2) $f \in V_j \Leftrightarrow f(2x) \in V_{j+1}$ (dilation invariance)

(1.3) $f \in V_0 \Leftrightarrow f(x-1) \in V_0$ (shift invariance).

(1.4) V_0 has a shift-invariant Riesz basis $(\varphi(x-k))_{k \in \mathbb{Z}}$.

The function φ is called a *scaling function* for (V_j).

Since $V_{-1} \subset V_0$, we may write $\varphi(\frac{x}{2})$ as a combination of the functions $\varphi(x-k)$:

(1.5) $$\varphi\left(\frac{x}{2}\right) = \sum_{k \in \mathbb{Z}} a_k \varphi(x-k) \quad \text{with} \quad (a_k)_{k \in \mathbb{Z}} \in \ell^2(\mathbb{Z}) \ .$$

Equivalently, we can write the Fourier transform $\hat{\varphi}$ of φ,

$$\hat{\varphi}(\xi) = \int \varphi(x) e^{-ix\xi} dx,$$

1991 *Mathematics Subject Classification*. Primary 46E20, 46E40, 47A58.

in the following way:

$$(1.6) \quad \hat{\varphi}(2\xi) = m_0(\xi)\hat{\varphi}(\xi) \text{ a.e. with } m_0(\xi) = \frac{1}{2}\sum_{k\in\mathbb{Z}} a_k e^{-ik\xi} \in L^2(\mathbb{R}/2\pi\mathbb{Z}).$$

If φ has enough decay at infinity ($| x |^{1/2+\epsilon} \varphi \in L^2$ for some $\epsilon > 0$), then $\hat{\varphi}$ is continuous and so is m_0 (since $m_0 \in H^{1/2+\epsilon}(\mathbb{R}/2\pi\mathbb{Z})$), so that (1.6) holds for every ξ and moreover:

$$(1.7) \qquad \hat{\varphi}(\xi) = \hat{\varphi}(0)\prod_{j=1}^{\infty} m_0\left(\frac{\xi}{2^j}\right) \quad \text{with} \quad \hat{\varphi}(0) \neq 0.$$

The function m_0 is then called a *scaling filter* for (V_j).

From now on, we will always assume that the scaling functions φ we will deal with are in $L^2(| x |^{1+2\epsilon} dx)$ for some $\epsilon > 0$, and that the associated scaling filters m_0 are in $H^{1/2+\epsilon}(\mathbb{R}/2\pi\mathbb{Z})$. For such scaling functions, we will speak of a *regular multiresolution analysis*.

Given two regular multiresolution analyses (V_j), (V_j^*), the following assertions are equivalent:

$$(1.8) \qquad L^2 = V_0 \oplus (V_0^*)^\perp$$

(1.9)   There is a bounded projection operator P_0 onL^2 such that

$$\operatorname{Ran} P_0 = V_0 \quad \text{and} \quad \operatorname{Ker} P_0 = (V_0^*)^\perp.$$

(1.10)   There are scaling funclions φ for (V_j) and φ^* for (V_j^*) such that:

$$\langle \varphi \mid \varphi^*(x-k)\rangle = \delta_{k,0}.$$

(1.11)   There are scaling filters m_0 for (V_j) and m_0^* for (V_j^*) such that:

$$m_0(\xi)\overline{m_0^*(\xi)} + m_0(\xi+\pi)\overline{m_0^*(\xi+\pi)} = 1.$$

We then speak of *dual scaling functions* φ, φ^*, of *dual scaling filters* m_0, m_0^* and (following A. Cohen, I. Daubechies and J. C. Feauveau [**7**]) of *bi-orthogonal multiresolution analyses*.

To the dual scaling function φ, φ^* (with associated filters m_0, m_0^*) we can moreover associate *dual wavelets* ψ, ψ^*, defined by:

$$(1.12) \qquad\qquad \hat{\psi}(\xi) = e^{-i\frac{\xi}{2}}\overline{m_0^*\left(\frac{\xi}{2}+\pi\right)}\hat{\varphi}\left(\frac{\xi}{2}\right)$$

$$(1.13) \qquad\qquad \hat{\psi}^*(\xi) = e^{-i\frac{\xi}{2}}\overline{m_0\left(\frac{\xi}{2}+\pi\right)}\hat{\varphi}^*\left(\frac{\xi}{2}\right).$$

The functions $\psi(x-k)$, $k \in \mathbb{Z}$, are then a Riesz basis for $W_0 = V_1 \cap V_0^{*\perp}$ and the $\psi^*(x-k)$, $k \in \mathbb{Z}$, are a Riesz basis for $W_0^* = V_1^* \cap V_0^\perp$; in addition,

$$(1.14) \qquad\qquad \langle \psi \mid \psi^*(x-k)\rangle = \delta_{k,0}.$$

As usual, we define $\psi_{j,k}$ and $\psi_{j,k}^*$ for $j \in \mathbb{Z}$ and $k \in \mathbb{Z}$ as:

$$(1.15) \qquad \qquad \psi_{j,k}(x) = 2^{j/2} \psi(2^j x - k)$$

$$(1.16) \qquad \qquad \psi_{j,k}^*(x) = 2^{j/2} \psi^*(2^j x - k) \,;$$

we have of course the bi-orthogonality relationship

$$(1.17) \qquad \qquad \langle \psi_{j,k} \mid \psi_{\ell,p}^* \rangle = \delta_{j,\ell} \delta_{k,p} \,.$$

If the scaling functions φ, φ^* belong to $L^2(\mid x \mid^{1+2\epsilon} dx)$ for some $\epsilon > 0$, then they also belong to the Sobolev space H^α for some $\alpha > 0$ (see by instance L. Hervé [11]); the associated wavelets ψ, ψ^* are then in $L^2(\mid x \mid^{1+2\epsilon} dx) \cap H^\alpha$ and it can be shown that the family $(\psi_{j,k})_{j \in \mathbb{Z}, k \in \mathbb{Z}}$ is a Riesz basis of $L^2(\mathbb{R})$ with the family $(\psi_{j,k}^*)$ as its dual basis.

More generally, a *wavelet basis* of $L^2(\mathbb{R})$ will be a Riesz basis $(\psi_{j,k})_{j \in \mathbb{Z}, k \in \mathbb{Z}}$ of $L^2(\mathbb{R})$ generated from one function ψ through the dyadic dilations and translations (1.15). There are wavelet bases for which the dual basis is not a wavelet basis; when the dual basis is a wavelet basis $(\psi_{j,k}^*)_{j \in \mathbb{Z}, k \in \mathbb{Z}}$, we will speak of *bi-orthogonal wavelet bases* $(\psi_{j,k})$, $(\psi_{j,k}^*)$ and of *dual wavelets* ψ, ψ^*.

The main object we will study in this paper will be the *large-scales projector* P_0 defined by:

$$(1.18) \qquad \qquad P_0 f = \sum_{j < 0} \sum_{k \in \mathbb{Z}} \langle f \mid \psi_{j,k}^* \rangle \psi_{j,k} \quad (f \in L^2(\mathbb{R})) \,.$$

As we shall see, many properties of bi-orthogonal wavelet bases and of bi-orthogonal multiresolution analyses can be expressed usefully in term of the projector P_0.

2. How to rewrite the projection operator $\mathbf{P_0}$ (and applications).

If we look at formula (1.18), we can see that this formula, of constant use in wavelet theory, has many shortcomings.

For instance, P_0 commutes with integer shifts:

$$(2.1) \qquad \text{for } k \in \mathbb{Z} \text{ and } f \in L^2, \quad P_0\{f(x - k)\} = \{P_0 f\}(x - k) \,,$$

even though none of the terms $\sum_{k \in \mathbb{Z}} \langle f \mid \psi_{j,k}^* \rangle \psi_{j,k}$ does (for $j < 0$). S. Mallat pointed out how inconvenient this can be (for stereo matching by instance); this was one of his reasons for proposing what he called the dyadic wavelet representation instead of wavelet bases [23].

The shift invariance of P_0 can be much more easily read off from the formula

$$(2.2) \qquad \qquad P_0 f = f - \sum_{j \geq 0} \sum_{k \in \mathbb{Z}} \langle f \mid \psi_{j,k}^* \rangle \psi_{j,k}$$

since now each term $\sum_{k \in \mathbb{Z}} \langle f \mid \psi_{j,k}^* \rangle \psi_{j,k}$ is shift-invariant (it commutes with shifts by multiples of $\frac{1}{2^j}$, hence with integer shifts). This feature will make formula (2.2) very useful below. However, it expresses $P_0 f$ as the superposition of smaller and smaller scales, hence of more and more irregular functions, which makes it less useful from other points of view.

When the wavelet bases are provided by bi-orthogonal multiresolution analyses (V_j), (V_j^*) with dual scaling functions φ, φ^*, then P_0 can be rewritten as:

$$(2.3) \qquad P_0 f = \sum_{k \in \mathbb{Z}} \langle f \mid \varphi^*(x - k) \rangle \varphi(x - k).$$

This is another way of writing our formulas (1.9) and (1.10); of course, this can be valid if and only if the functions φ and φ^* exist; we will address this problem in a moment.

There is still another way of writing P_0. Let τ_h be the shift operator $\tau_h f = f(x - h)$, Q_j be the projection operator $Q_j f = \sum_{k \in \mathbb{Z}} \langle f \mid \psi_{j,k}^* \rangle \psi_{j,k}$. We have:

$$(2.4) \qquad \text{for} \quad k \in \mathbb{Z}, \qquad \tau_k P_0 = P_0 \tau_k$$

$$(2.5) \qquad \text{for} \quad k \in \mathbb{Z} \text{ and } j \in \mathbb{Z}, \qquad \tau_{k2^{-j}} Q_j = Q_j \tau_{k2^{-j}},$$

hence

$$(2.6) \quad P_0 f = \sum_{j=-N}^{-1} 2^j \sum_{k=0}^{2^{-j}-1} \tau_k Q_j \tau_{-k} f + \frac{1}{2^N} \sum_{k=0}^{2^N-1} \tau_k \left(\sum_{j=-\infty}^{-N-1} Q_j \right) \tau_{-k} f.$$

Now, we can prove quite easily that for any $f \in L^2$, the term

$$\frac{1}{2^N} \sum_{k=0}^{2^N-1} \tau_k \left(\sum_{j=-\infty}^{-N-1} Q_j \right) \tau_{-k} f$$

goes to 0 in L^2 as N goes to $+\infty$, so that:

$$(2.7) \qquad P_0 f = \sum_{j<0} 2^j \sum_{k=0}^{2^{-j}-1} (\tau_k Q_j \tau_{-k}) f$$

which can also be rewritten as:

$$(2.8) \qquad P_0 f = \sum_{j<0} \sum_{k \in \mathbb{Z}} \langle f \mid 2^j \psi^*(2^j(x - k)) \rangle 2^j \psi(2^j(x - k)).$$

Formula (2.8) can be read as a discrete version of Mallat's dyadic wavelet representation. This can also be usefully compared to the result by C. K. Chui and X. Shi that every dyadic tight wavelet frame leads to a dyadic wavelet representation [4].

We have now four formulas for computing $P_0 f$: (1.18), (2.2), (2.3) and (2.8). The point is: how can we use them, and which one should we use? Formula (2.3) is known to be very useful, since it is the foundation for the fast wavelet transform of S. Mallat; if we want to compute the wavelet coefficients $d_{j,k} = \langle f \mid \psi_{j,k}^* \rangle$ for

$k \in \mathbb{Z}$ and $j_0 \leq j \leq j_1$, it can be performed quickly by computing first the scaled coefficients $s_{j_1+1,k} = \langle f \mid \varphi^*_{j_1+1,k} \rangle$, $k \in \mathbb{Z}$, and then computing $(s_{j,k})_{k \in \mathbb{Z}}$ and $(d_{j,k})_{k \in \mathbb{Z}}$ from $(s_{j+1,k})_{k \in \mathbb{Z}}$.

Formula (2.8) should be also useful, providing an over-sampled but shift-invariant wavelet representation for functions $f \in \mathrm{Ran}\, P_0$. The ways of getting rid of redundancy in this representation should then be explored in the way S. Mallat proposed for the dyadic wavelet representation.

But formula (2.8) has other applications. As a matter of fact, I introduced it two years ago to prove that every orthonormal wavelet basis $(\psi_{j,k})$ with a compactly supported or exponentially decaying wavelet ψ is provided by a multiresolution analysis (i.e. the $\psi(x-k)$, $k \in \mathbb{Z}$, are a Riesz basis of $W_0 = V_1 \cap V_0^{\perp}$ for some multiresolution analysis (V_j)) and that the associated space V_0 has an orthonormal basis $\varphi(x-k)$ with a compactly supported or exponentially decaying scaling function φ [19], [21]. The assumptions on the decay of ψ ensured that $\hat{\psi}$ was analytical. Recently, P. Auscher showed that the proof could be extended to (sufficiently) continuous $\hat{\psi}$; he also obtained a very nice result about the Hardy space $H^{(2)}$ of analytical signals [1], [3].

THEOREM 2.1. Let $(\psi_{j,k})_{j \in \mathbb{Z}, k \in \mathbb{Z}}$ be an orthonormal wavelet basis of $L^2(\mathbb{R})$ such that the wavelet ψ belongs to $L^2(\mid x \mid^{1+2\epsilon} dx) \cap H^{\alpha}$ for some positive ϵ and α. Then the space $V_0 = \mathrm{Ran}\, P_0$ has an orthonormal basis $\varphi(x-k)$, $k \in \mathbb{Z}$, with $\varphi \in L^2(\mid x \mid^{1+2\epsilon} dx)$. Moreover φ can be chosen with compact support if ψ has a compact support, or exponentially decaying if ψ is, or rapidly decaying if ψ is.

THEOREM 2.2 (P. AUSCHER). The Hardy space $H^{(2)} = \{f \in L^2 / \mathrm{Supp}\, \hat{f} \subset [0, +\infty)\}$ has no orthonormal wavelet basis $(\psi_{j,k})_{j \in \mathbb{Z}, k \in \mathbb{Z}}$ with a wavelet ψ in $L^2(\mid x \mid^{1+2\epsilon} dx) \cap H^{\alpha}$ for some positive ϵ and α.

The proof is very simple. If $(\psi_{j,k})$ is an orthonormal wavelet basis of L^2 or $H^{(2)}$, we may define the projector P_0 by (1.18); this projector can be rewritten as in (2.8), or, due to the Poisson summation formula:

$$(2.9) \qquad \widehat{P_0 f}(\xi) = \sum_{j<0} \sum_{k \in \mathbb{Z}} \hat{f}(\xi + 2k\pi) \overline{\hat{\psi}(2^{-j}(\xi + 2k\pi))} \hat{\psi}(2^{-j}\xi) \quad \text{a.e.}$$

which gives, for all $f \in \mathrm{Ran}\, P_0$,

$$(2.10) \quad \sum_{k \in \mathbb{Z}} \mid \hat{f}(\xi + 2k\pi) \mid^2 = \sum_{j<0} \left| \sum_{k \in \mathbb{Z}} \hat{f}(\xi + 2k\pi) \overline{\hat{\psi}(2^{-j}(\xi + 2k\pi))} \right|^2 \quad \text{a.e.}$$

By the Cauchy-Schwarz inequality, we get, for all $f \in \mathrm{Ran}\, P_0$,

$$(2.11)$$
$$\sum_{k \in \mathbb{Z}} \mid \hat{f}(\xi + 2k\pi) \mid^2 \leq \sum_{k \in \mathbb{Z}} \mid \hat{f}(\xi + 2k\pi) \mid^2 \sum_{j<0} \sum_{k \in \mathbb{Z}} \mid \hat{\psi}(2^{-j}(\xi + 2k\pi)) \mid^2 \quad .$$

Choose f to be $f = 2^j \psi(2^j \xi)$ $(j < 0)$; then summing over j gives:

$$(2.12) \quad \sum_{j<0} \sum_{k \in \mathbb{Z}} \mid \hat{\psi}(2^{-j}(\xi + 2k\pi)) \mid^2 \leq \left(\sum_{j<0} \sum_{k \in \mathbb{Z}} \mid \hat{\psi}(2^{-j}(\xi + 2k\pi)) \mid^2 \right)^2 \quad \text{a.e.}$$

Let $\eta(\xi) = \sum_{j<0} \sum_{k \in \mathbb{Z}} \mid \hat{\psi}(2^{-j}(\xi + 2k\pi)) \mid^2$; we have seen that $\eta(\xi) \leq \eta(\xi)^2$, so that for almost all ξ we have $\eta(\xi) = 0$ or $\eta(\xi) \geq 1$. Now, if $\psi \in L^2(\mid x \mid^{1+2\epsilon} dx) \cap H^\alpha$, the series which defines η converges uniformly on every compact of $(0, 2\pi)$, so that η is continuous on $(0, 2\pi)$; moreover

$$\frac{1}{2\pi} \int_0^{2\pi} \eta(\xi) d\xi = \frac{1}{2\pi} \sum_{j<0} 2^j \parallel \hat{\psi} \parallel_2^2 = 1 .$$

Hence $\eta \geq 1$ on $(0, 2\pi)$ (since η cannot be identically 0) and therefore $\eta = 1$ on $(0, 2\pi)$. Now (2.12) is an equality instead of an inequality, which means that all the Cauchy-Schwarz inequalities were equalities; therefore all the vectors in $\ell^2(\mathbb{Z})$, $e_j(\xi)$ defined by $e_j(\xi) = (\hat{\psi}(2^{-j}(\xi + 2k\pi)))_{k \in \mathbb{Z}}$ are, for fixed ξ, co-linear when j goes through $-\mathbb{N}^*$, i.e. there exist functions γ_j and α on $(0, 2\pi)$ such that $\hat{\psi}(2^{-j}(\xi + 2k\pi)) = \gamma_j(\xi)\alpha(\xi + 2k\pi)$. By 2π-periodic extension of the γ_j, α we can assume this to be true a.e. for ξ in $\mathbb{R}$. Since $\eta = 1$, it follows that $\Gamma(\xi) = \sum_{j<0} \mid \gamma_j(\xi) \mid^2$ and $A(\xi) = \sum_k \mid \alpha(\xi + 2k\pi) \mid^2$ satisfy $\Gamma(\xi)A(\xi) = 1$. By replacing, if necessary, γ_j and α by $\gamma_j \Gamma^{-1/2}$ and $\alpha A^{-1/2}$ we may as well assume that $\sum_{j<0} \mid \gamma_j(\xi) \mid^2 = 1$. Substituting into (2.9) then leads to

$$\widehat{P_0 f}(\xi) = \sum_{k \in \mathbb{Z}} \hat{f}(\xi + 2k\pi)\overline{\alpha(\xi + 2k\pi)} \, \alpha(\xi)$$

or

$$P_0 f = \sum_{k \in \mathbb{Z}} \langle f, \varphi(x - k)\rangle \varphi(x - k) ,$$

where $\hat{\varphi} = \alpha$. Since $\parallel \varphi \parallel^2 = \frac{1}{2\pi}\parallel \alpha \parallel^2 = \frac{1}{2\pi}\int_0^{2\pi} A(\xi) d\xi = 1$, it follows that the $\varphi(x - k)$, $k \in \mathbb{Z}$, are an orthonormal basis of Ran P_0. In the case of Theorem 2.1 (a wavelet basis in L^2), we just have to prove that we may choose this scaling function φ in $L^2(\mid x \mid^{1+2\epsilon} dx)$ (or with compact support, or with exponential decay); this point will be discussed later. In the case of Theorem 2.2 (a basis in $H^{(2)}$), we write $\hat{\varphi}(2\xi) = m_0(\xi)\hat{\varphi}(\xi)$ for some $m_0 \in L^2(\mathbb{R}/2\pi\mathbb{Z})$; it is easy to see that we must have:

$$(2.13) \qquad\qquad \mid m_0(\xi) \mid^2 + \mid m_0(\xi + \pi) \mid^2 = 1$$

$$(2.14) \qquad\qquad \mid m_0(\xi + \pi) \parallel \hat{\varphi}(\xi) \mid = \mid \hat{\psi}(2\xi) \mid$$

$$(2.15) \qquad\qquad \mid m_0(\xi) \mid^2 = \sum_{k \in \mathbb{Z}} \mid \hat{\psi}(2\xi + 4k\pi) \mid^2 .$$

Now $\mu(\xi) = |m_0(\xi)|^2$ is a Hölderian function (because of (2.15) and $\psi \in L^2$ $(|x|^{1+2\epsilon} dx)$); moreover we have for all N:

$$(2.16) \qquad |\hat{\varphi}(\xi)|^2 \leq \prod_{j=1}^{N} \mu\left(\frac{\xi}{2^j}\right).$$

If $\mu(0) < 1$, then we obtain $\hat{\varphi} = 0$ which is absurd; therefore $\mu(0) = 1$ and we obtain:

$$(2.17) \qquad |\hat{\varphi}(\xi)|^2 \leq \Phi(\xi) = \prod_{j=1}^{\infty} \mu\left(\frac{\xi}{2^j}\right),$$

with

$$(2.18) \qquad \int_{-\infty}^{+\infty} \Phi(\xi)d\xi \leq 2\pi \quad (\text{since } \mu \text{ satisfies } (2.13)),$$

whence

$$(2.19) \qquad \|\hat{\varphi}\|_2^2 \leq 2\pi - \int_{-\infty}^{0} \Phi(\xi)d\xi < 2\pi$$

since $\Phi(0) = 1$ and Φ is continuous. It follows that we cannot have $\|\varphi\|_2 = 1$, leading to a contradiction. This proves Theorem 2.2.

There is another proof of Theorem 2.1, using (1.18) and (2.2) instead of (2.8). This proof can be easily extended to bi-orthogonal wavelet bases so that we have the following theorem:

THEOREM 2.3. Let $(\psi_{j,k})$, $(\psi_{j,k}^*)$ be bi-orthogonal wavelet bases of $L^2(\mathbb{R})$ such that ψ and ψ^* belong to $L^2(|x|^{1+2\epsilon} dx) \cap H^\alpha$ for some positive ϵ and α. Then the spaces $V_0 = \operatorname{Ran} P_0$ and $V_0^* = (\operatorname{Ker} P_0)^\perp$ generate bi-orthogonal multiresolution analyses and the dual scaling functions φ, φ^* can be chosen in $L^2(|x|^{1+2\epsilon} dx)$.

Moreover φ, φ^* can be chosen with compact supports if ψ, ψ^* are compactly supported, or chosen with exponential decay if ψ and ψ^* are exponentially decaying.

(This extension to the biorthogonal case was first carried out for compactly supported biorthogonal wavelets in [15]; the general case was announced in [3] and [20], and proved – by different methods – in [2] and [16].)

Let us first note that Theorem 2.3 implies Theorem 2.1; as a matter of fact, if $V_0 = V_0^*$ and $\psi = \psi^*$, we can by Theorem 2.3 exhibit a Riesz basis of $V_0(\varphi(x - k))_{k \in \mathbb{Z}}$ with $\varphi \in L^2(|x|^{1+2\epsilon} dx)$. Then the Gram orthonormalization $\varphi \to \tilde{\varphi}$, where $\tilde{\varphi}$ is defined by its Fourier transform

$$\hat{\tilde{\varphi}}(\xi) = \frac{\hat{\varphi}(\xi)}{\left(\sum_{k \in \mathbb{Z}} |\hat{\varphi}(\xi + 2k\pi)|^2\right)^{1/2}},$$

gives us an orthonormal basis $(\tilde{\varphi}(x - k))$ of V_0 with $\tilde{\varphi}$ in $L^2(|x|^{1+2\epsilon} dx)$ as well. In a similar way, orthonormalizing a Riesz basis $(\varphi(x-k))$ with an exponentially decaying scaling function φ gives us an exponentially decaying orthonormal scaling function $\tilde{\varphi}$. Last, if $V_0 = V_0^*$ and φ, φ^* are compactly supported dual scaling functions, then necessarily $\varphi = \lambda \varphi^*$ with a positive scalar λ, so that $\frac{1}{\sqrt{\lambda}} \varphi$ is the required compactly supported orthonormal scaling function.

The proof of Theorem 2.3 relies on a simple estimate on the kernel of the projection operator P_0. We first define Q_0 by:

$$(2.20) \qquad Q_0 f = \sum_{k \in \mathbb{Z}} \langle f \mid \psi_{0,k}^* \rangle \psi_{0,k} \,;$$

this is an integral operator $Q_0 f = \int q(x,y) f(y) dy$, with kernel

$$(2.21) \qquad q(x,y) = \sum_{k \in \mathbb{Z}} \overline{\psi^*(y-k)} \psi(x-k) \,.$$

It is now easy to see that if ψ, ψ^* are in $L^2(\mid x \mid^{1+2\epsilon} dx)$, then

$$(2.22) \qquad \int \int_{0 \leq x \leq 1} \mid x-y \mid^{1+2\epsilon} (\mid q(x,y) \mid^2 + \mid q(y,x) \mid)^2 dx dy < +\infty$$

while, since ψ, ψ^* are also in H^α,

$$(2.23) \qquad q(x,y) \in L_{\text{loc}}^{r_0} \quad \text{for some} \quad r_0 > 2 \,.$$

Combining this with (1.18) shows that P_0 is an integral operator, with kernel

$$p(x,y) = \sum_{j<0} 2^j q(2^j x, 2^j y) \,;$$

because of (2.23), $p(x,y)$ is locally square-integrable. On the other hand, (2.2) implies that outside from the diagonal $x = y$ we have also

$$p(x,y) = - \sum_{j \geq 0} 2^j q(2^j x, 2^j y) \,;$$

by (2.22) the integral

$$\int \int_{0 \leq x \leq 1} \mid x-y \mid^{1+2\epsilon} (\mid p(x,y) \mid^2 + \mid p(y,x) \mid^2) dx dy$$

is finite. Theorem 2.3 can now be reduced to a corollary of the following theorem:

THEOREM 2.4. Let P_0 be a bounded projection operator on $L^2(\mathbb{R})$ so that:
* P_0 is invariant under integer shifts;
* P_0 is an integral operator with kernel $p(x,y)$ satisfying:

$$(2.24) \qquad \int \int_{0 \leq x \leq 1} (\mid p(x,y) \mid^2 + \mid p(y,x) \mid^2)(1+ \mid x-y \mid)^{1+2\epsilon} < +\infty$$

for some positive ϵ. Then $V_0 = \operatorname{Ran} P_0$ has a shift-invariant Riesz basis

$$(\varphi_\delta(x-k))_{1 \leq \delta \leq D, k \in \mathbb{Z}} \,.$$

The φ_δ can be chosen in $L^2(\mid x \mid^{1+2\epsilon} dx)$ and the dual basis

$$(\varphi_\delta^*(x - k))_{1 \leq \delta \leq D, k \in \mathbb{Z}}$$

in $V_0^* = (\operatorname{Ker} P_0)^\perp$ satisfies then $\varphi_\delta^* \in L^2(\mid x \mid^{1+2\epsilon} dx)$.

As a matter of fact, there is a very easy characterization of functions $\varphi_1, \dots, \varphi_D$ generating a Riesz basis $(\varphi_\delta(x - k))$ of V_0. For $f_1, \dots, f_N \in L^2(\mathbb{R})$, let $M(f_1, \dots, f_N)$ be the *auto-correlation matrix* of $f_1, \dots, f_N$ defined by its coefficients

$$C_{i,j}(\xi) = \sum_{k \in \mathbb{Z}} \hat{f}_i(\xi + 2k\pi) \overline{\hat{f}_j}(\xi + 2k\pi) \,.$$

The matrix $M(f_1, \dots, f_N)$ belongs to the space $M_N(L^1(0, 2\pi))$ of the $N \times N$ matrices with coefficients in $L^1(0, 2\pi)$. Now, we have the following characterization:

LEMMA 2.5. The functions $\varphi_1, \dots, \varphi_D$ generate a Riesz basis $(\varphi_\delta(x - k))_{1 \leq \delta \leq D, k \in \mathbb{Z}}$ of V_0 if and only if they satisfy:
   (i) $M(\varphi_1, \dots, \varphi_D) \in M_D(L^\infty(0, 2\pi))$ (i.e. its coefficients are essentially bounded functions);
   (ii) $M(\varphi_1, \dots, \varphi_D)$ is invertible in $M_D(L^\infty(0, 2\pi))$;
   (iii) $\forall f \in V_0, \det(M(\varphi_1, \dots, \varphi_D, f)) = 0$ a.e.

In order to find $\varphi_1, \dots, \varphi_D$ with $M(\varphi_1, \dots, \varphi_D)$ invertible for all $\xi \in [0, 2\pi]$, we may begin by looking for local solutions (i.e. $M(\varphi_1, \dots, \varphi_D)$ invertible for any given $\xi_0 \in [0, 2\pi]$).

LEMMA 2.6. a) Let $\xi \in [0, 2\pi]$. For $f \in V_0 \cap L^2(\mid x \mid^{1+2\epsilon} dx)$, let f_ξ be defined by

$$f_\xi(x) = \sum_{k \in \mathbb{Z}} e^{-i\xi(x-k)} f(x - k) \,.$$

Then $f_\xi \in L^2([0, 1])$. Moreover if $\varphi_1, \dots, \varphi_d \in V_0 \cap L^2(\mid x \mid^{1+2\epsilon} dx)$ then the matrix $M(\varphi_1, \dots, \varphi_d)$ computed at the point ξ is the Gram matrix of the functions $\varphi_{1,\xi}, \dots, \varphi_{d,\xi}$.
   b) Let $V_\xi = \{f_\xi \mid f \in V_0 \cap L^2(\mid x \mid^{1+2\epsilon} dx)\}$. Under the hypotheses of Theorem 2.4, we have $\dim V_\xi < +\infty$. Moreover this dimension does not depend on ξ.

Lemma 2.6 is easy, because we know the projection operator of $L^2([0, 1])$ on V_ξ. It is the integral operator with kernel p_ξ given by:

$$(2.25) \qquad p_\xi(x, y) = \sum_{k \in \mathbb{Z}} e^{i\xi(y - k - x)} p(x, y - k) \,.$$

The estimate (2.24) ensures that $p_\xi \in L^2([0,1] \times [0,1])$; thus the projection operator on V_ξ is a Hilbert-Schmidt operator and V_ξ is finite-dimensional. Moreover, we can easily compute this dimension:

$$(2.26) \qquad \dim V_\xi = \int\int_{[0,1]\times[0,1]} p_\xi(x,y)p_\xi(y,x)dx\,dy$$

and we find that $\dim V_\xi$ is a continuous function of ξ, hence a constant function.

The proof of Theorem 2.4 now follows easily. We may find for any $\xi_0 \in [0,2\pi]$ a solution $M(\varphi_1^{\xi_0},\dots,\varphi_D^{\xi_0})$ invertible in a neighborhood of ξ_0; some easy topological arguments give us a global solution.

Theorem 2.3 is now also easy to prove. We must have $D = 1$ in case of a wavelet basis, since the space V_1 will have two shift-invariant Riesz bases, the basis

$$(\varphi_\delta(x-k))_{1\leq\delta\leq D,k\in\mathbb{Z}} \cup (\psi(x-k))_{k\in\mathbb{Z}}$$

and the basis

$$(\varphi_\delta(2x-2k))_{1\leq\delta\leq D,k\in\mathbb{Z}} \cup (\varphi_\delta(2x-2k-1))_{1\leq\delta\leq D,k\in\mathbb{Z}};$$

Lemma 2.5 gives us then $D + 1 = 2D$, hence $D = 1$.

The case of exponentially decaying wavelets is handled in exactly the same way (we obtain an exponential control of $p(x,y)$ when moving away from the diagonal). The case of compactly supported wavelets needs a further step. We first notice, due to formula (2.2), that $p(x,y)$ is properly supported, i.e. that for every compactly supported function f, $P_0 f$ and $P_0^* f$ are compactly supported. Now, we use the following result of [14]:

LEMMA 2.7. If (V_j) is a multiresolution analysis such that V_0 contains non trivial compactly supported functions, then there exists a scaling function φ in V_0 such that:
  * φ is compactly supported;
  * every compactly supported function in V_0 is a finite linear combination of the $\varphi(x-k)$, $k \in \mathbb{Z}$.
Moreover there exists a compactly supported $h \in L^2(\mathbb{R})$ such that $\langle \varphi \mid h(x-k) \rangle = \delta_{k,0}$.

We thus obtain compactly supported dual scaling functions φ, φ^*, where $\varphi^* = P_0^* h$. Theorem 2.3 is proved.

3. Hierarchical bases and wavelets.

Let us first recall some properties of *finitely generated shift-invariant spaces* in $L^2(\mathbb{R})$ [9], [18]. A finitely generated shift-invariant space, or *F.I.S. space*, will be a closed subspace V of $L^2(\mathbb{R})$ such that there is a finite set $\{g_1,\dots,g_L\}$ of compactly supported functions such that V is the closed linear span of the $g_\ell(x-k)$, $1 \leq \ell \leq L$, $k \in \mathbb{Z}$.

A F.I.S. space always has a shift-invariant Riesz basis $(\varphi_\delta(x-k))_{1\leq\delta\leq D, k\in\mathbb{Z}}$ with compactly supported basic functions φ_δ. The number D doesn't depend on the choice of the Riesz basis (see Lemma 2.5) and is called the length of V, written $D = \operatorname{len} V$.

A *fundamental set* $(\varphi_1, \ldots, \varphi_D)$ in V is a set of compactly supported functions such that the $\varphi_\delta(x-k)$, $1 \leq \delta \leq D$, $k \in \mathbb{Z}$, are a Riesz basis of V and such that each compactly supported function in V can be written as a finite linear combination of the $\varphi_\delta(x-k)$. Lemma 2.7 can be extended into the following lemma:

LEMMA 3.1. Let V be a F.I.S. space of length D. Then:
  (i) V has fundamental sets.
  (ii) A family $(\varphi_1, \ldots, \varphi_D)$ of D functions in V with compact support, is a fundamental set if and only if there exist D compactly supported functions $h_1, \ldots, h_D$ in $L^2(\mathbb{R})$ such that

$$(3.1) \qquad \langle h_i(x) \mid \varphi_j(x-k) \rangle = \delta_{i,j}\delta_{k,0}$$

The family $(h_1, \ldots, h_D)$ is called a dual family of $(\varphi_1, \ldots, \varphi_D)$.

If $(\varphi_1, \ldots, \varphi_D)$ is a fundamental set of V with a dual family $(h_1, \ldots, h_D)$, we have a projection operator from L^2 onto V defined by:

$$(3.2) \qquad Pf = \sum_{i=1}^{D}\sum_{k\in\mathbb{Z}} \langle f \mid h_i(x-k) \rangle \varphi_i(x-k).$$

This projection, which depends on the choice of the h_i, is called a *fundamental projection* on V. Theorem 2.4 can also be written as:

THEOREM 3.2. Let P be a bounded projection operator on $L^2(\mathbb{R})$. Then $V = \operatorname{Ran} P$ is a F.I.S. space and P is a fundamental projection on V if and only if:
  * P commutes with integer shifts;
  * P is an integral operator, the kernel of which is a properly supported locally square-integrable function.
If moreover P is self-adjoint ($V = (\operatorname{Ker} P)^{\perp}$), then V has an orthonormal basis $(\varphi_i(x-k))_{1\leq i\leq V, k\in\mathbb{Z}}$ with compactly supported basic functions φ_i.

Let now (V_j) be a multiresolution analysis such that V_0 contains non trivial compactly supported functions. Then V_0 is a F.I.S. space of length 1; we call φ a fundamental scaling function for V_0. The idea of the construction of a hierarchical basis for V_j (following Yserentand [26]) is to add to the function φ a function ψ such that (φ, ψ) is a fundamental set for V_1. We then have for $j \geq 0$ two bases for V_j: the "nodal" one, which is the basis $(2^{j/2}\varphi(2^j x - k))_{k\in\mathbb{Z}}$, and the "hierarchical" one, i.e. the basis

$$(\varphi(x-k))_{k\in\mathbb{Z}} \cup (2^{\ell/2}\psi(2^\ell x - k))_{0\leq\ell\leq j-1, k\in\mathbb{Z}}.$$

The algorithms for going from one basis to the other one are analogous to the F.W.T. of S. Mallat (with F.I.R. filters) and we have for $v_0 \in V_0$ and $w_\ell \in W_\ell$ (where W_ℓ is the closed linear span of the $\psi_{\ell,k}$, $k \in \mathbb{Z}$):

$$(3.3) \quad C_1(j) \left(\| v_0 \|_2^2 + \sum_{\ell=0}^{j-1} \| w_\ell \|_2^2 \right)^{1/2} \leq \| v_0 + \sum_{\ell=0}^{j-1} w_\ell \|_2$$

$$\leq C_2(j) \left(\| v_0 \|_2^2 + \sum_{\ell=0}^{j-1} \| w_\ell \|_2^2 \right)^{1/2}.$$

THEOREM 3.3.

(i) The set (φ, ψ) is fundamental in V_1 if and only if the trigonometric polynomial $m_1(\xi)$ such that $\hat{\psi}(2\xi) = m_1(\xi)\hat{\varphi}(\xi)$ and the scaling filter m_0 associated to φ satisfy:

(3.4)
$$\exists \lambda \in \mathbb{C}^*, \quad \exists N \in \mathbb{Z}, \quad m_0(\xi)m_1(\xi + \pi) - m_0(\xi + \pi)m_1(\xi) = \lambda e^{i(2N+1)\xi}.$$

(ii) $C_2(j)$ in formula (3.3) can be bounded independently of j if and only if $\int \psi \, dx = 0$.

(iii) Let $\int \psi \, dx$ be 0. Then $C_1(j)^{-1}$ can be bounded independently of j if and only if there exists a multiresolution analysis (V_j^*) with a compactly supported scaling function φ^* such that φ, φ^* are dual scaling functions and that $W_0 = V_1 \cap (V_0^*)^\perp$. The scaling filter m_0^* associated to φ^* is then $m_0^* = \frac{1}{\lambda} e^{+i(2N+1)\xi} \bar{m}_1(\xi + \pi)$.

Thus, one-dimensional stable hierarchical bases are just bi-orthogonal wavelet bases. This is, I believe, a very significant (though essentially obvious) result. It means that, in order to construct a hierarchical basis for $P1$ or $Q1$ finite elements on a bounded polygonal domain in $\mathbb{R}^2$, we should try to understand what should be a wavelet basis in such a framework. A first attempt has been made in [13].

Another consequence of this theorem is that if we want to construct an efficient multigrid theory for finite elements (such as cubic elements associated to Hermitian interpolation) we should try to construct generalized bi-orthogonal multiresolution analyses (where the spaces V_0 and V_0^* are now F.S.I. spaces of length greater than one), as introduced by many authors (see [11] by instance).

4. A lexicon for compactly supported dual scaling functions.

Let (V_j), (V_j^*) be bi-orthogonal multiresolution analyses with compactly supported dual scaling functions φ, φ^*. Then one can define (V_j) and (V_j^*) by any of the following four objects:

(i) the compactly supported dual scaling functions φ, φ^* (they are unique up to scalar multiplication and integer shift: $\varphi \to \lambda \varphi(x - k)$, $\varphi^* \to \frac{1}{\lambda} \varphi^*(x - k)$, $\lambda \in \mathbb{C}^*$, $k \in \mathbb{Z}$)

(ii) the dual F.I.R. scaling filters m_0, m_0^* (they are unique up to a multiplication by a character $m_0 \to e^{-ik\xi}m_0$, $m_0^* \to e^{-ik\xi}m_0^*$, $k \in \mathbb{Z}$)

(iii) the compactly supported dual wavelets ψ, ψ^* (they are unique up to scalar multiplication and integer shift)

(iv) the fundamental large-scales projection operator P_0.

In this section, I would like to describe the properties of the bi-orthogonal multiresolution analyses in each of the four languages we may deal with; we obtain equivalent descriptions but different interpretations. The three problems we will address are: i) the generation of bi-orthogonal multiresolution analyses; ii) the Strang-Fix conditions; iii) derivation and finite differences in wavelet theory.

4.1. The generation of bi-orthogonal multiresolution analysis. Let us first recall some results on scaling functions and on scaling filters.

PROPOSITION 4.1. Let φ be a compactly supported square-integrable function on $\mathbb{R}$. Then φ is a scaling function (for some multiresolution analysis (V_j)) if and only if

i)
$$\inf_{\xi \in \mathbb{R}} \sum_{k \in \mathbb{Z}} |\hat{\varphi}(\xi + 2k\pi)|^2 > 0;$$

ii) there exists $m_0 \in \mathcal{C}^\infty(\mathbb{R}/2\pi\mathbb{Z})$ such that $\hat{\varphi}(2\xi) = m_0(\xi)\hat{\varphi}(\xi)$.

PROPOSITION 4.2 (A. COHEN [**5**]). Let m_0 be a trigonometric polynomial. Then m_0 is a scaling filter (for some multiresolution analysis (V_j) with compactly supported scaling function) if and only if:

(i) $m_0(0) = 1$;

(ii) $\prod_{j=1}^{\infty} m_0(\frac{\xi}{2^j})$ is square-integrable;

(iii) there exists a compact set K which can be written as a finite union of disjoint closed interval such that:

(j) $\sum_{k \in \mathbb{Z}} \chi_K(\xi + 2k\pi) = 1$ a.e.

(jj) $\forall \xi \in K, \forall j \in \mathbb{N}^*$, $m_0(\frac{\xi}{2^j}) \neq 0$.

Condition (iii) can be rewritten as:

(iii') m_0 satisfies:

(j') $|m_0(\xi)|^2 + |m_0(\xi + \pi)|^2 > 0$

(jj') $\forall \xi_0 \in (0, 2\pi)$, $\exists N \in \mathbb{N}$, $m_0(2^N\xi_0 + \pi) \neq 0$.

Condition (ii) can be rewritten in the following way:

PROPOSITION 4.3 ([**6**], [**11**], [**25**]). Let m_0 satisfy $m_0(0) = 1$ and Cohen's criterion (iii) of Proposition 4.2. Then $\prod_{j=1}^{\infty} m_0\left(\frac{\xi}{2^j}\right)$ belongs to L^2 if and only if:

  (k) $m_0(\pi) = 0$;

  (kk)   the operator $T_0 : f \to \mid m_0(\frac{\xi}{2}) \mid^2 f(\frac{\xi}{2}) + \mid m_0(\frac{\xi}{2} + \pi) \mid^2 f(\frac{\xi}{2} + \pi)$, where $\mid m_0(\xi) \mid^2 = \sum_{k=-N}^{N} \alpha_\ell e^{-i\ell\xi}$, restricted to

$$F_N = \left\{ f = \sum_{-N}^{N} \beta_{\ell\ell} e^{-i\ell\xi}, \quad (\beta_\ell) \in \mathbb{C}^{2N+1}/f(0) = 0 \right\},$$

has a spectral radius less than 1 $(\rho(T_0 \mid_{F_N}) < 1)$.

The following theorem gives four different ways to characterize or construct a bi-orthogonal multiresolution analysis with compactly supported scaling functions, based on respectively the scaling functions, the scaling filters, the wavelets or the projectors.

THEOREM 4.4.

  (i) Let φ, φ^* be compactly supported square-integrable functions. Then they are dual scaling functions if and only if they satisfy:

    * the two-scale equations

$$(4.1) \; \varphi\left(\frac{x}{2}\right) = \sum_{k=k_0}^{k_1} a_k \varphi(x - k) \text{ for some finite sequence } (a_k)$$

    (4.2)

$$\varphi^*\left(\frac{x}{2}\right) = \sum_{k=k_2}^{k_3} b_k \varphi^*(x - k) \text{ for some finite sequence } (b_k)$$

    * the bi-orthogonality relationship:

$$(4.3) \qquad\qquad \langle \varphi(x) \mid \varphi^*(x - k) \rangle = \delta_{k,0}.$$

  (ii) Let m_0, m_0^* be trigonometric polynomials. Then they are dual scaling filters if and only if they satisfy:

    * Albert Cohen's condition i), ii), iii) of Proposition 4.2;

    * the duality relationship:

$$(4.4) \qquad m_0(\xi)\bar{m}_0^*(\xi) + m_0(\xi + \pi)\bar{m}_0^*(\xi + \pi) = 1.$$

  (iii) Let ψ, ψ^* be compactly supported square-integrable functions such that ψ and ψ^* belong to H^α for some positive α. Then they are dual wavelets (associated to bi-orthogonal multiresolution analyses) if and only if they satisfy:

    * the bi-orthogonality relationship:

$$(4.5) \qquad\qquad \langle \psi_{j,k} \mid \psi_{\ell,p}^* \rangle = \delta_{j,\ell}\delta_{k,p}$$

* the completeness requirement: L^2 is the closed linear span of the $\psi_{j,k}$ $(j \in \mathbb{Z}, k \in \mathbb{Z})$.

(iv) Let P_0 be a bounded projection operator on $L^2(\mathbb{R})$. Then it is a fundamental large-scale projector (for bi-orthogonal multiresolution analyses with compactly supported dual scaling functions) if and only if:

* P_0 commutes with integer shifts:

$$(4.6) \qquad P_0\{f(x-k)\} = \{P_0f\}(x-k).$$

* P_0 commutes with $D_{-2}P_0$ and P_0D_2 (where $D_2f = f(2x)$ and $D_{-2}f = f(\frac{x}{2})$) : $P_0D_{-2}P_0 = D_{-2}P_0$ and $P_0D_2P_0 = P_0D_2$.
* P_0 is an integral operator $P_0f(x) = \int p_0(x,y)f(y)dy$ where:
    - p_0 is locally square-integrable;
    - p_0 is properly supported:

$$(4.7) \qquad \exists M > 0, \quad |x-y| > M \Rightarrow p_0(x,y) = 0.$$

* P_0 satisfies the dimension requirement:

(4.8)

$$\int\int_{[0,1]\times[0,1]} \left(\sum_{k\in\mathbb{Z}} p_0(x,y-k)\right)\left(\sum_{k\in\mathbb{Z}} p_0(y,x-k)\right) dxdy = 1.$$

4.2. The Strang-Fix conditions. The Strang-Fix conditions have been introduced in the early 70's to describe the approximation properties of some shift-invariant operators [10]. In wavelet theory, they can be written as follows:

THEOREM 4.5. Let (V_j), (V_j^*) be bi-orthogonal multiresolution analysis with compactly supported dual scaling functions. Then for any $N \geq 0$ the following assertions are equivalent:

(i) The projection operator P_0 satisfies

$$(4.9) \quad \exists C > 0, \quad \forall f \in H^{N+1}, \quad \| f - P_0f \|_2 \leq C \| f^{(N+1)} \|_2 .$$

(ii) The projection operators P_j from L^2 on V_j in direction of $(V_j^*)^\perp$ satisfy:

$$(4.10) \qquad \forall f \in H^N, \lim_{j\to+\infty} 2^{jN} \| f - P_jf \|_2 = 0.$$

(iii) The projection operator P_0 keeps the low-degreed polynomials invariant:

$$(4.11) \qquad \forall p \in \{0,1,\dots,N\} \quad P_0(x^p) = x^p .$$

(iv) The dual scaling functions φ, φ^* satisfy:

$$(4.12) \quad \forall p \in \{0,1,\dots,N\} \quad \sum_{k\in\mathbb{Z}}(x-k)^p\varphi(x-k) = (-i)^p\hat{\varphi}^{(p)}(0)$$

(and no condition on φ^*).

(v) The dual scaling filters m_0, m_0^* satisfy

$$(4.13) \qquad \forall p \in \{0, 1, \ldots, N\} \quad \left(\frac{d}{d\xi}\right)^p m_0(\pi) = 0$$

(and no conditions on m_0^*).
(vi) The dual wavelets ψ, ψ^* satisfy:

$$(4.14) \qquad \forall p \in \{0, 1, \ldots, N\} \quad \int x^p \psi^* dx = 0$$

(and no conditions on ψ).

(4.9) to (4.11) are the celebrated Strang-Fix conditions. (4.13) is the sum rule of I. Daubechies and (4.14) is the usual oscillation requirement for an analyzing wavelet.

4.3. Derivation and finite differences in wavelet theory. Let $(\psi_{j,\ell})$, $(\psi_{j,k}^*)$ be bi-orthogonal wavelet bases with compactly supported wavelets ψ, ψ^*. Let us assume that ψ is moreover in the Sobolev space H^1 (then it is in some space $H^{1+\epsilon}$ with $\epsilon > 0$). Then we obtain a new bi-orthogonal system by deriving ψ and integrating $\psi^* : (\tilde{\psi}_{j,k}), (\tilde{\psi}_{j,k}^*)$ with $\tilde{\psi} = \tilde{\psi}'$ and $\tilde{\psi}^* = \int_x^{+\infty} \psi^*(t)dt$. These new basis are related to a new bi-orthogonal multiresolution analysis, and the derivation can therefore be described in each language.

THEOREM 4.6. Let $(V_j), (V_j^*)$ bi-orthogonal multiresolution analyses with compactly supported dual scaling functions. Let $V_0 \subset H^1$ and define $\tilde{V}_j$ as the closure in L^2 of $\{g \in L^2 / \exists f \in V_j, \quad g = f'\}$ and $\tilde{V}_j^*$ as $\tilde{V}_j^* = \{g \in H^1 / g' \in V_j^*\}$. Then $(\tilde{V}_j), (\tilde{V}_j^*)$ are still bi-orthogonal multiresolution analyses with compactly supported dual scaling functions and we have:
(i) the dual scaling functions φ, φ^* satisfy:

$$(4.15) \quad \frac{d\varphi}{dx} = \tilde{\varphi}(x) - \tilde{\varphi}(x-1) \quad \text{and} \quad \int_x^{x+1} \varphi^*(t)dt = \tilde{\varphi}^*(x)$$

where $\tilde{\varphi}$ and $\tilde{\varphi}^*$ are compactly supported dual scaling functions for $(\tilde{V}_j)$ and $(\tilde{V}_j^*)$;
(ii) the dual scaling filters $\tilde{m}_0$ and $\tilde{m}_0^*$ associated to $\tilde{\varphi}$ and $\tilde{\varphi}^*$ are given by:

$$(4.16) \quad \tilde{m}_0(\xi) = \frac{2}{1 + e^{-i\xi}} m_0(\xi) \quad \text{and} \quad \tilde{m}_0^*(\xi) = \frac{e^{i\xi} + 1}{2} m_0^*(\xi)$$

where m_0 and m_0^* are the filters associated to φ and φ^*.
(iii) The dual wavelets ψ and ψ^* satisfy

$$(4.17) \qquad \psi' = \tilde{\psi} \quad \text{and} \quad \int_x^{+\infty} \psi^*(t)dt = \tilde{\psi}^*(x)$$

where $\tilde{\psi}, \tilde{\psi}^*$ are dual compactly supported wavelets for $(\tilde{V}_j), (\tilde{V}_j^*)$.

(iv) The projector P_0 V_0 and the projector $\tilde{P}_0$ on $\tilde{V}_0$ satisfy:

$$(4.18) \qquad \forall f \in H^1, \quad \frac{d}{dx}(P_0 f) = \tilde{P}_0 \left(\frac{df}{dx} \right) .$$

Formulas (4.15) and (4.16) were noticed by G. Malgouyres in 1990. They have been also used very elegantly by L. Villemoes in [**25**]. Formula (4.17) has been of constant use in harmonic analysis for continuous wavelets, for which this is a very natural statement to say that the derivative or the primitive of a wavelet is a wavelet. Formula (4.18) expresses the very interesting fact that the derivative of the approximation is an approximation of the derivative. I have shown in [**12**] and [**17**] how to use this property to construct easily wavelet basis for divergence free vector functions.

5. Conclusion.

There are by now two great books on wavelet theory, the book by Y. Meyer [**24**] mainly dealing with the scaling functions point of view, and the book by I. Daubechies [**8**] mainly dealing with the filter point of view. I tried to stress here a third point of view, the operator point of view; it is my belief that it will help us to investigate new fields in wavelet theory, beyond the problem of existence of scaling functions solved in Theorems 2.1 and 2.3.

REFERENCES

1. P. Auscher, *Il n'existe pas de bases d'ondelettes régulières dans l'espace de Hardy $H^2(\mathbb{R})$*, C. R. Acad. Sci. Paris **315** (1992), 769–772, série 1.
2. _____, *Solution of two problems on wavelets*, preprint, 1992.
3. _____, *Toute base d'ondelettes régulières de $L^2(\mathbb{R})$ est issue d'une analyse multirésolution régulière*, C. R. Acad. Sci. Paris **315** (1992), 1227–1230, série 1.
4. C. K. Chui and X. Shi, *Inequalities of Littlewood-Paley for frames and wavelets*, SIAM J. Math. Anal. **24** (1993), 263–277.
5. A. Cohen, *Ondelettes, analyses multirésolutions et traitement numérique du signal*, Ph.D. thesis, University of Paris IX, 1990.
6. A. Cohen and I. Daubechies, *A stability criterion for bi-orthogonal wavelet bases and their related subband coding scheme*, Duke Math. J. **68** (1992), 313–335.
7. A. Cohen, I. Daubechies, and J. C. Feauveau, *Biorthogonal bases of compactly supported wavelets*, Comm. Pure & Appl. Math. **45** (1992), 485–560.
8. I. Daubechies, *Ten lectures on wavelets*, SIAM, 1992, CBMS Lecture Notes nr. 61.
9. C. de Boor, R. A. De Vore, and A. Ron, *The structure of finitely generated shift-invariant spaces in $L^2(\mathbb{R}^d)$*, preprint, 1992.
10. G. Fix and G. Strang, *Fourier analysis of the finite element method in Ritz-Galerkin theory*, Studia Appl. Math. **48** (1969), 265–273.
11. L. Herve, *Méthodes d'opérateurs quasi-compacts en analyse multirésolution, applications à la construction de bases d'ondelettes et à l'interpolation*, Ph.D. thesis, Univ. of Rennes I, 1992.
12. A. Jouini and P. G. Lemarie-Rieusset, *Analyses multirésolutions bi-orthogonales sur l'intervalle et applications*, to appear in Ann. I.H.P. (Analyses non linéaire).
13. _____, *Ondelettes sur un ouvert borné du plan*, preprint, 1992.
14. P. G. Lemarie, *Fonctions à support compact dans les analyses multirésolutions*, Revista Matematica Iberoamericana **7** (1991), 157–182.

15. P. G. Lemarie-Rieusset, *Ondelettes généralisées et fonctions d'échelle à support compact*, to appear in Revista Matematica Iberoamericana.

16. _____, *Projecteurs invariants, matrices de dilatation, ondelettes et analyses multirésolutions*, submitted to Revista Matematica Iberoamericana.

17. _____, *Analyses multirésolutions non orthogonales, commutation entre projecteurs et dérivation et ondelettes vecteurs à divergence nulle*, Revista Matematica Iberoamericana **8** (1992), 221–237.

18. _____, *Eléments finis 1d sur grilles régulières, bases hiérarchiques et théorie des ondelettes*, preprint, 1992.

19. _____, *Existence de fonctions-père pour les ondelettes à support compact*, C. R. Acad. Sci. Paris **314** (1992), 17–19.

20. _____, *Fonctions d'échelle pour les ondelettes de dimension n*, C. R. Acad. Sci. Paris **316** (1993), 145–148.

21. _____, *Sur l'existence des analyses multirésolutions en théorie des ondelettes*, Revista Matematica Iberoamericana **8** (1993), 457–474.

22. S. Mallat, *A theory for multiresolution signal decomposition: the wavelet representation*, IEEE Trans. PAM **11** (1989), 674–693.

23. _____, *Zero-crossings of a wavelet transform*, IEEE Trans. Inf. Th. **37** (1991), 1019–1033.

24. Y. Meyer, *Ondelettes et opérateurs, I: Ondelettes*, Hermann, Paris, 1990.

25. L. Villemoes, *A wavelet analysis of the two-scale dilation equations*, preprint, 1992.

26. H. Yserentand, *On the multilevel splitting of finite element spaces*, Numer. Math. **49** (1986), 379–412.

UNIVERSITÉ DE PARIS-SUD, MATHÉMATIQUE, BÂT. 425, 91405 ORSAY, FRANCE

E-mail: uanh007@frors31.bitnet

Proceedings of Symposia in Applied Mathematics
Volume **47**, 1993

Wavelets and Differential Operators

PHILIPPE TCHAMITCHIAN

Notations

We begin with the following preliminary set of definitions, which may differ slightly from those in the other chapters.

Bases of wavelets are denoted $(\psi_\lambda)_{\lambda \in \Lambda}$. In dimension 1, the set of indices Λ consists of all the dyadic rationals $\lambda = (k + \frac{1}{2})2^{-j}$, $j, k \in \mathbb{Z}$. In dimension n, it consists of the points $\lambda = (k + \frac{\epsilon}{2})2^{-j}$, $j \in \mathbb{Z}$, $k \in \mathbb{Z}^n$, $\epsilon = (\epsilon_1, \ldots, \epsilon_n)$ with $\epsilon_i = 0$ or 1, and $\epsilon \neq (0, \ldots, 0)$. As often as possible, we write j, k instead of $j(\lambda), k(\lambda)$, and simply forget ϵ. We define $\psi_\lambda(x) = 2^{jn/2} \psi_{\frac{\epsilon}{2}}(2^j x - k)$, where we assume for simplicity that each $\psi_{\frac{\epsilon}{2}}$ is in $\mathcal{S}(\mathbb{R}^n)$ and has all its moments vanishing, i.e. $\int \psi_{\frac{\epsilon}{2}}(x) \, x_1^{\ell_1} \ldots x_n^{\ell_n} \, d^n x = 0$ for $\ell_1, \ldots, \ell_n \in \mathbb{N}$. We shall also describe these as the "cancellation properties" of the $\psi_{\frac{\epsilon}{2}}$. These bases come from a multi-resolution analysis (MRA) $(V_j)_{j \in \mathbb{Z}}$, and a basis of V_j is given by the set of functions $(\varphi_{jk})_{k \in \mathbb{Z}^n}$, where $\varphi \in \mathcal{S}(\mathbb{R}^n)$ and $\varphi_{jk}(x) = 2^{jn/2} \varphi(2^j x - k)$. All the prerequisites on wavelets are in Ingrid Daubechies' lecture.

Except for the letter ψ, indexing a set of functions by Λ, like in $(\theta_\lambda)_{\lambda \in \Lambda}$, has the usual meaning, i.e. $\theta_\lambda(x)$ does *not* stand for $2^{jn/2} \theta_{\frac{\epsilon}{2}} (2^j x - k)$ in general: if we intend the latter, we will write $(\theta)_\lambda$ instead of θ_λ.

1. Wavelets and Fourier: differential operators with constant coefficients.

When faced with an orthonormal basis of wavelets, it is natural to wonder (and this question is often asked) whether they are eigenfunctions for some operator, like complex exponentials are for the laplacian. One easily checks that no standard differential operator admits wavelets as eigenfunctions. However, this is not a complete answer to our question, and we shall come back to this below. But let us reverse our point of view for a while, and ask how the laplacian (or any elliptic operator with constant coefficients) acts on wavelets?

1991 *Mathematics Subject Classification.* Primary 35A35, 35A40, 47A58.

The direct action is easily written:

$$(1.1) \qquad \Delta(\psi_\lambda) = 4^j (\Delta\psi)_\lambda \; ,$$

where we have used only the translation invariance and the homogeneity of the operator. In other words, if $\theta = \Delta\psi$, then the laplacian can be decomposed into the product of two operators:

$$(1.2) \qquad \Delta = T\Lambda^2 \; ,$$

where Λ is defined by $\Lambda(\psi_\lambda) = 2^j \psi_\lambda$, and T by $T(\psi_\lambda) = (\theta)_\lambda$.

What can be said about the set of functions $((\theta)_\lambda)$?

Because the (ψ_λ) are a basis, the $(\theta)_\lambda$ constitute a dense set in $L^2(\mathbb{R}^n)$. Moreover, if we define $\tilde\theta_{\frac{\epsilon}{2}} = \Delta^{-1}(\psi_{\frac{\epsilon}{2}})$, which is a nice function thanks to the cancellation properties of $\psi_{\frac{\epsilon}{2}}$, then it is easy to see that the $(\tilde\theta)_\lambda$ are biorthogonal to the $(\theta)_\lambda$:

$$(1.3) \qquad \langle (\theta)_\lambda, \; (\tilde\theta)_\mu \rangle = \delta_{\lambda\mu} \; .$$

Finally, $((\theta)_\lambda)$ and $((\tilde\theta)_\lambda)$ from two biorthogonal unconditional basis of L^2 if we have

$$\forall f \in L^2 \qquad \int |f|^2 \; \sim \; \sum_\lambda |\langle f, (\theta)_\lambda \rangle|^2$$

$$\sim \; \sum_\lambda |\langle f, (\tilde\theta)_\lambda \rangle|^2 \; .$$

($\sim$ stands for equivalence between two norms.) By duality, using (1.3), these inequalities are equivalent to the existence of $c > 0$ such that

$$(1.4) \qquad \forall f \in L^2 \qquad \sum_\lambda |\langle f, (\theta)_\lambda \rangle|^2 \le c \int |f|^2$$

and

$$(1.5) \qquad \forall f \in L^2 \qquad \sum_\lambda |\langle f, (\tilde\theta)_\lambda \rangle|^2 \le c \int |f|^2 \; .$$

One can prove that these inequalities hold, thanks again to the cancellation properties of $(\theta)_\lambda$ and $(\tilde\theta)_\lambda$, together with their good localization and smoothness. Let us take this result for granted for now; we will come back to this in §5 below, where we will prove a more general result that implies (1.4) and (1.5).

Formulas (1.1) and (1.2) also lead to the following two results. First, we have at our disposal two factorizations of the inverse of the laplacian:

$$\Delta^{-1} = \Lambda^{-2} T^{-1}$$

(we know that T is invertible) and

$$\Delta^{-1} = \tilde T \Lambda^{-2} \; ,$$

where by definition, $\tilde{T}(\psi_\lambda) = (\tilde{\theta})_\lambda$. Second, we can use (1.1) to derive a characterization of the Sobolev space $H^2(\mathbb{R}^n)$. For $f = \sum_\lambda c_\lambda \psi_\lambda \in H^2$, we have indeed

$$\Delta f = \sum_\lambda 4^j c_\lambda (\theta)_\lambda = T\left(\sum_\lambda 4^j c_\lambda \psi_\lambda\right),$$

so that

$$\int |\Delta f|^2 \sim \sum_\lambda 4^{2j} |c_\lambda|^2.$$

The reader can easily check that the same kind of results hold for all the homogeneous elliptic constant coefficient operators, as well as for all their functional powers. In particular, the norms in the Sobolev spaces $H^s(\mathbb{R}^n)$ are equivalent to

$$\left(\sum_\lambda (1 + 4^{js}) |\langle f, \psi_\lambda\rangle|^2\right)^{1/2},$$

for every $s \in \mathbb{R}$.

Comparing with the Fourier expansion, one sees that the extra operator T (or $\tilde{T}$) makes the whole difference between $\Delta(\psi_\lambda) = 4^j(\theta)_\lambda$ and $\Delta(e^{i\omega.x}) = -|\omega|^2 e^{i\omega.x}$. This operator is the price to pay when using the wavelet transform instead of the Fourier transform to analyze the laplacian. Two advantages we obtain are a strong functional stability and a good numerical efficiency. The latter will be illustrated further; the reader is also referred to Gregory Beylkin's lecture. The former is a consequence of the fact that (ψ_λ) is an unconditional basis of $L^p(\mathbb{R}^n)$ if $1 < p < +\infty$, and that the operator T is still continuous and invertible on these spaces. As a consequence, one can characterize $W_p^2(\mathbb{R}^n)$ and, more generally, $W_p^s(\mathbb{R}^n)$ for every $s \in \mathbb{R}$. The same idea works as well for the Besov spaces.

To conclude this paragraph, let us point out that the formula

$$\Delta = T\Lambda^2$$

is nothing but the wavelet version of Calderón's factorization of differential operators ([4]). In order to see how it works in the variable coefficients case, we now consider the pointwise multiplication operator, starting with multiplication by relatively smooth functions.

2. Multiplication by lipschitz functions.

We have just seen how the laplacian can be decomposed in a nice way with respect to a wavelet basis. A similar nice decomposition exists for the pointwise multiplication with (not necessarily very) smooth functions; the key point here is the localization of the wavelets. A typical example is that of the multiplication by $a(x)$, a bounded and lipschitz function on $\mathbb{R}^n$.

If ψ_λ is a small scale wavelet, i.e. if j is positive and large, then ψ_λ is very much concentrated around the point λ, and it seems reasonable to approximate the product $a\psi_\lambda$ by $a(\lambda)\psi_\lambda$. It therefore makes sense to introduce the operator P_a defined by

$$P_a(\psi_\lambda) = a(\lambda)\psi_\lambda \ .$$

Then M_a, the pointwise multiplication by a, is given by

$$M_a = P_a + R \ ,$$

where R is regularizing of order 1, i.e. continuous from L^2 to H^1.

To prove this, compute the image of a wavelet:

$$R\psi_\lambda = [a - a(\lambda)] \ \psi_\lambda \ ,$$

so that

$$
\begin{aligned}
\nabla(R\psi_\lambda) &= (\nabla a) \ \psi_\lambda \\
&\quad + [a - a(\lambda)] \ 2^j \ (\nabla\psi)_\lambda \ .
\end{aligned}
$$

It follows that, say, $\partial_1 R$ is L^2-continuous if and only if U is, where U is the operator defined by

(2.1) $$U(\psi_\lambda) = \sigma_\lambda \ ,$$

with

$$\sigma_\lambda(x) = [a(x) - a(\lambda)] \ 2^j \ (\partial_1\psi)_\lambda(x) \ .$$

The σ_λ have some regularity: they are lipschitz with $\| \nabla \sigma_\lambda \|_\infty \le C2^j \ \| \nabla a \|_\infty$, uniformly in $\lambda \in \Lambda$. Also, because

$$2^j |a(x) - a(\lambda)| \le \| \nabla a \|_\infty \ |2^j(x - \lambda)| \ ,$$

the σ_λ are rapidly decreasing, satisfying the estimates

$$|\sigma_\lambda(x)| \le C_N \| \nabla a \|_\infty \ 2^{jn/2} \ (1 + |2^j x - k|)^{-N}$$

for every $N \in \mathbb{N}$, uniformly in λ. It follows that, insofar as localization is concerned, the σ_λ behave like wavelets, even though they are not obtained by dilation and translation from a single function. This means that the operator U is comparable to the operators T and $\tilde{T}$ we met before. We will see that U is also continuous on L^2, thus giving the regularizing property of R.

Because a regularizing operator need not be a small operator, this result is not directly applicable to numerical applications. But, using the structure of multi-resolution analysis, it is easy to modify the preceding decomposition in a suitable way. Choose an index $p \in \mathbb{Z}$, and take as orthonormal basis the set of functions $(\varphi_{pk}, \ \psi_\lambda)_{j \ge p, \ k \in \mathbb{Z}^n}$. Then, define $\tilde{P}_a$ and $\tilde{R}$ (depending on p) by

$$
\begin{aligned}
\tilde{P}_a(\varphi_{pk}) &= a(k2^{-p}) \ \varphi_{pk} \\
\tilde{P}_a(\psi_\lambda) &= a(\lambda) \ \psi_\lambda \\
\tilde{R} &= M_a - \tilde{P}_a \ .
\end{aligned}
$$

The same method as before leads to the result that $\tilde{R}$ is still regularizing of order 1, and, as an operator on L^2, has a norm controlled by $C\,2^{-p}\|\bigtriangledown a\|_\infty$, where C is a constant independent of a and p.

The operators P_a and $\tilde{P}_a$, called paraproducts, belong to a class of operators that have been used by J. M. Bony to study the propagation of singularities in nonlinear equations. They depend on the choice of the wavelet basis. For a fixed wavelet basis, they are diagonal operators, like the operator Λ we previously defined. The wavelets turn out therefore to be eigenfunctions of Λ (which is a pseudo-differential operator) and of paraproducts.

3. Example and numerical application.

The next step in this study is combining differential operators and pointwise multiplication with lipschitz functions. We restrict ourselves to the following relevant example:

$$L = -\operatorname{div} A \bigtriangledown +1 \;,$$

where $A = A(x)$ is a bounded and lipschitz matrix-valued function. As usual, we assume the existence of $\delta > 0$ such that

$$\forall x, \xi \in \mathbb{R}^n \qquad A(x)\xi \cdot \bar{\xi} \geq \delta|\xi|^2 \;.$$

We denote by B the bilinear form

$$B(f,g) = \int A \bigtriangledown f \cdot \bigtriangledown \bar{g} + \int f\bar{g} \;.$$

By the Lax-Milgram theorem, every equation

$$Lf = g \;,$$

where $g \in L^2$, has a unique solution f, which is here in H^2. To compute f, classical algorithms, like finite element methods, start with a Galerkin approximation: instead of solving

$$B(f,\varphi) = \int g\bar{\varphi}$$

for every φ in H^1, they solve the restricted problem

$$B(f_V,\varphi) = \int g\bar{\varphi}$$

for every φ in a subspace V which has been chosen as the approximation space, and in which the computed f_V lies. It is well-known that this leads to ill-conditioned problems if V is associated to a fine grid, and this is why multigrid methods, for example, have been developed.

An alternative approach is possible: by making use of the preceding ideas, one can construct the inverse of L, in a way which is not only constructive, but algorithmic. To explain this, let us fix $p \in \mathbb{Z}$, and denote by $G_p : V_p \to V_p$

the Galerkin operator, whose matrix in the basis $(\varphi_{pk})_{k \in \mathbb{Z}^n}$ is the inverse of $(B(\varphi_{pk}, \varphi_{p\ell}))_{k,\ell}$. If $j \geq p$, we define θ_λ by the equation

$$-\mathrm{div}\, A(\lambda) \bigtriangledown \theta_\lambda = \psi_\lambda \; ;$$

we also define an operator P by

$$P(\psi_\lambda) = \theta_\lambda \; ,$$

acting on $V_p^\perp$. If we denote by π_p, $\pi_p^\perp$ the orthonormal projections onto V_p, $V_p^\perp$, and if η_p is the natural embedding of V_p into L^2, then the operator

$$A_p = \eta_p\, G_p\, \pi_p + P\pi_p^\perp$$

is an approximation of L^{-1}. In fact, the following result holds [10]: if the multi-resolution analysis is suitably chosen, and if we define U_p by

$$(3.1) \qquad\qquad L(\eta_p\, G_p\, \pi_p + P\pi_p^\perp) = I - U_p \; ,$$

then the operator U_p is continuous on L^2 and has a spectral radius bounded by $C\, 2^{-p}$, where C depends on A and on the wavelet choice, but not on p. Hence, if p is large enough, one has

$$L^{-1} = A_p \sum_{k=0}^{+\infty} U_p^k \; ,$$

and

$$A_p \sum_{k=0}^{K} U_p^k = L^{-1} - L^{-1}U_p^{K+1} \; .$$

Let us illustrate this by solving the equation

$$-\frac{d}{dx}\, \theta(x)\, \frac{d}{dx} f(x) + f(x) = g(x)$$

on $[0,1]$ with periodic boundary condition, and $\theta(x) = 2 + \cos 2\pi x$; for this illustration, we choose g so that $f(x) = \sin 2\pi x$ is the exact solution.

In [8], the approximate solution $f_n^{(p)}$ in V_8 is computed, using the above formula for L^{-1}, with different values of $p \leq 7$, and different values for the summation index cut-off n in the series $\sum U_p^k$. On the other hand, this approximate solution in V_8 is also computed via the classical Galerkin method. In both cases, one then compares with the true projection of f onto V_8. Figure 1 plots the ratio $\frac{\|f_n^{(p)} - f\|_2}{\|f\|_2}$ for different values of $p \leq 7$. For $p \geq 5$ this ratio plateaus at $E_8 < 10^{-7}$ after very few (3 or 4) iterations. More details, including a comparison with the Galerkin method, and other examples can be found in [8].

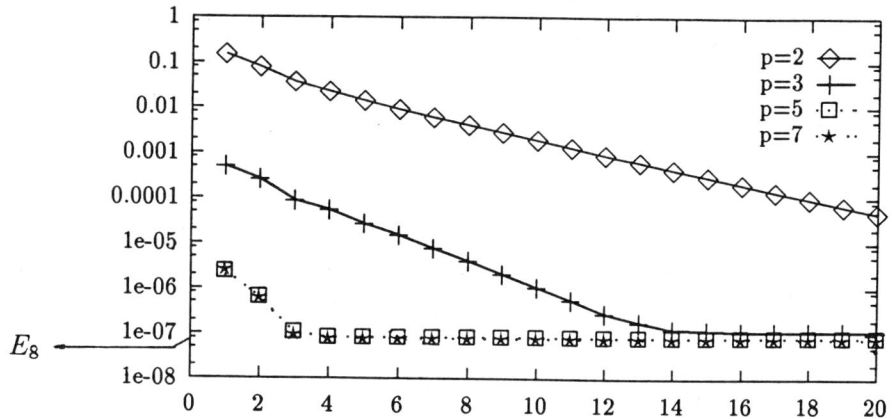

FIGURE 1. Evolution of $\frac{\|f_n^{(p)} - f\|_2}{\|f\|_2}$ as a function of the number n of iterations, for different values of p.

4. Multiplication by rough functions and operators with non-smooth coefficients.

We now turn our attention to the multiplication by non-smooth functions. In such a case, the approximation by a paraproduct can still be defined, replacing the pointwise values $a(\lambda)$ by appropriate mean values, for example $2^{jn/2} \langle a, \varphi_{jk} \rangle$; but the remainder will then not be regularizing. At the present time, the problem of analyzing an operator like $-\operatorname{div} A \triangledown$, where A is bounded and elliptic, but without any regularity assumption on A, is wide open, at least in the general case.

Let us be a little more concrete about what we would like to know. The operator is defined through the bilinear form B:

$$\forall f, g \in H^1(\mathbb{R}^n) \qquad B(f, g) = \int A \triangledown f \cdot \triangledown g \,,$$

and its domain, dense in L^2, is an exotic functional space. The problems we address are:

   (i) find an unconditional basis of the domain of the operator;
   (ii) find an unconditional basis of $H^1(\mathbb{R}^n)$ orthonormal with respect to the form B.

In dimension 1, these two problems are completely solved for $L = D a D$ ($D = \frac{1}{i} \frac{d}{dx}$), where a is bounded and accretive (i.e. $\operatorname{Re} a \geq \delta > 0$ for a given δ), and with L defined on any open set of $\mathbb{R}$, with any kind of linear boundary condition ([1],[2]). Let us explain how this can be done. For the sake of simplicity, we restrict ourselves to the case where the variable x ranges over all of $\mathbb{R}$, i.e. L acts on a dense subspace of $L^2(\mathbb{R})$.

The key result is the existence of an unconditional basis of $L^2(\mathbb{R})$, $(\theta_\lambda)_{\lambda \in \Lambda}$, satisfying good estimates, and orthonormal with respect to the accretive bilinear form

$$b(f, g) = \int f(x)\, g(x)\, \frac{1}{a(x)}\, dx \ .$$

The estimates we have in mind are

(4.1) $$|\theta_\lambda(x)| \leq C_N\, 2^{j/2}\, (1 + |2^j x - k|)^{-N}$$

for every integer N, uniformly in λ, and

(4.2) $$\|(\theta_\lambda)'\|_\infty \leq C\, 2^{j/2}\, 2^j \ ,$$

uniformly in λ as well. Moreover, these basis functions satisfy cancellation properties relative to the measure $\frac{1}{a(x)}\, dx$:

$$\forall m \in \mathbb{N} \ \forall \lambda \in \Lambda \ \int x^m\, \theta_\lambda(x)\, \frac{1}{a(x)} dx = 0 \ .$$

Taking the (θ_λ) for granted (we shall come back to their existence below), we now define τ_λ and σ_λ by

$$\begin{aligned}
\tau_\lambda &= 2^j D^{-1}\left(\frac{1}{a}\,\theta_\lambda\right) , \\
\sigma_\lambda &= 2^{-j} D\left(\theta_\lambda\right) .
\end{aligned}$$

(The existence of D^{-1} is the specific feature of dimension 1 that we use here in a crucial way.) By construction, $L\tau_\lambda = 4^j \sigma_\lambda$, and $B(\tau_\lambda, \tau_\mu) = \delta_{\lambda\mu}$. The (τ_λ) and the (σ_λ) each constitute biorthogonal unconditional bases of $L^2(\mathbb{R})$, and the (τ_λ) also characterize $H^1(\mathbb{R})$ and Dom (L). We can indeed write any f in $L^2(\mathbb{R})$ as

$$f = \sum_\lambda c_\lambda\, \tau_\lambda \ ,$$

with $c_\lambda = \int f \sigma_\lambda$; these characterization statements then amount to

$$\|f\|_{L^2} \ \sim \ \left(\sum_\lambda |c_\lambda|^2\right)^{1/2} ,$$

$$\|f\|_{H^1} \ \sim \ \left(\sum_\lambda (1 + 4^j)\, |c_\lambda|^2\right)^{1/2} ,$$

$$\|f\|_{Dom\ (L)} \ \sim \ \left(\sum_\lambda (1 + 4^{2j})\, |c_\lambda|^2\right)^{1/2} .$$

These three formulas can be proved with little effort once we know (θ_λ). First, using the relations $D\tau_\lambda = 2^j \frac{1}{a}\, \theta_\lambda$ and $L\tau_\lambda = 4^j \sigma_\lambda$, they are reduced to

$$\int |f|^2 \sim \sum_\lambda \left|\int f\sigma_\lambda\right|^2 \sim \sum_\lambda \left|\int f\tau_\lambda\right|^2 .$$

Then, thanks to biorthogonality, we can reduce this to

$$(4.3) \qquad \sum_\lambda \left| \int f \sigma_\lambda \right|^2 \le C \int |f|^2$$

and

$$(4.4) \qquad \sum_\lambda \left| \int f \tau_\lambda \right|^2 \le C \int |f|^2 \ .$$

Finally, these inequalities are proved in the same way as (1.4) and (1.5). (See below.)

Consider now the proof of the existence of the θ_λ. There are two main steps. The first is the effective construction of the family, the second is the proof that it is an unconditional basis of $L^2(\mathbb{R})$.

The first step proceeds by mimicking the usual construction of (ψ_λ), but without using the Fourier transform. Starting from $V_j \subset V_{j+1}$, one defines

$$X_j = \{ f \in V_{j+1}; \ b(f, V_j) = 0 \} \ .$$

Then, $V_{j+1} = V_j \oplus X_j$. To obtain a basis of X_j, one begins with the construction of a basis of V_j, orthonormal with respect to the form b. This is done by computing the square-root inverse of the matrix

$$(b \, (\varphi_{jk}, \ \varphi_{j\ell}))_{k,\ell} \ ,$$

and proving estimates on its elements (fast decay off the diagonal), uniformly in j. Next, one uses the projection from V_{j+1} onto X_j, parallel to V_j, which is an isomorphism between W_j (the usual orthogonal complement of V_j in V_{j+1}) and X_j, and hence maps $(\psi_\lambda)_{j(\lambda)=j}$ onto a basis of X_j. Finally, an orthonormalization procedure, like in V_j, gives the θ_λ, $j(\lambda) = j$.

The whole construction relies on the following ingredients.

 i) The functional calculus in a good class of matrices. The matrices $(b \, (\varphi_{jk}, \ \varphi_{j\ell}))_{k,\ell}$ satisfy good estimates of the type

$$\forall N \in \mathbb{N}, \ \exists C_N \quad \text{so that} \quad |m_{k\ell}| \le C_N (1 + |k - \ell|)^{-N} \ .$$

Note that this class of matrices is stable under taking the inverse and taking the square-root ([7]).

 ii) Good conditioning: the matrices $(b \, (\varphi_{jk}, \ \varphi_{j\ell}))_{k,\ell}$ are well conditioned, i.e. their condition numbers are uniformly bounded for $j \in \mathbb{Z}$. This is the missing point when one tries to generalize this construction to $H^1(\mathbb{R}^n)$, with the bilinear form B replacing b.

The second step is the proof of unconditionality. Formally, every f in L^2 can be written as

$$f = \sum_\lambda b(f, \theta_\lambda) \, \theta_\lambda \ ,$$

and one has to prove

$$\int |f|^2 \sim \sum_\lambda |b\,(f,\theta_\lambda)|^2 \sim \sum_\lambda \left|\int f\theta_\lambda\right|^2 \,.$$

Once again, this reduces to the inequality

(4.5) $$\sum_\lambda \left|\int f\theta_\lambda\right|^2 \leq C \int |f|^2 \,.$$

We now turn to the basic tool used to prove this and previous similar inequalities.

5. A useful class of operators.

We call $\mathcal{C}$ the class of operators T defined by

$$T(\psi_\lambda) = \theta_\lambda \,,$$

where (θ_λ) is any set of functions satisfying the estimates (4.1) and (4.2).

The operator T will be a bounded (or continuous) operator in L^2 if and only if

$$\left\|\sum_\lambda c_\lambda \theta_\lambda\right\|_{L^2} \leq C \left(\sum_\lambda |c_\lambda|^2\right)^{1/2} \,,$$

which, by duality, is equivalent to

(5.1) $$\left(\sum_\lambda \left|\int f\,\theta_\lambda\right|^2\right)^{1/2} \leq C\,\|f\|_{L^2} \,.$$

We met operators belonging to $\mathcal{C}$ and inequalities like (5.1) at each crucial step:
- the factorization of Δ and the inequalities (1.4) and (1.5);
- the regularizing property of the remainder $R = M_a - P_a$ and the continuity of U, defined in (2.1);
- the estimate on the spectral radius of V_p in (3.1), which also uses the class $\mathcal{C}$;
- the analysis of $L = D\,a\,D$, in (4.3) and (4.4),
- and finally the estimate (5.1).

If T is in $\mathcal{C}$, then its kernel can be expanded as

$$K(x,y) = \sum_\lambda \theta_\lambda(x)\,\overline{\psi}_k(y) \,,$$

and the estimates on θ_λ give, if $x \neq y$,

$$|K(x,y)| \leq \frac{C}{|x-y|^n} \,,$$

$$|\nabla_x K(x,y)| + |\nabla_y K(x,y)| \leq \frac{C}{|x-y|^{n+1}}$$

(in fact, we would have to modify (4.2) a little, but this is a minor point without any consequence). Such a kernel is said to be of Calderón-Zygmund type ([**5**], [**9**]). If T is continuous on L^2, T is said to be a Calderón-Zygmund operator. The classical theory of Calderón-Zygmund operators asserts that they are continuous on L^p, $1 < p < \infty$, from the real Hardy space H^1 to L^1, and from L^∞ to BMO: this is why one can characterize L^p, H^1 and BMO with a wavelet basis.

Hence, the central problem, given an operator T in the class $\mathcal{C}$, is to know whether it is continuous on L^2 or not. This question is completely solved by the following result ([**5**], [**6**], [**9**]): an operator T in the class $\mathcal{C}$ is continuous on L^2 if and only if the numbers $\int \theta_\lambda$ satisfy the Carleson condition:

$$(5.2) \qquad \exists C > 0, \text{ such that for any cube } Q, \qquad \sum_{Q_\lambda \subset Q} \left| \int \theta_\lambda \right|^2 \leq C|Q| \,,$$

where Q_λ is the cube centered in λ and of volume 2^{-jn}. The proof of this characterization is based upon two concepts from real analysis, the maximal function and Carleson's measure.

Proving inequalities (1.4), (1.5), (4.3) and (4.4) is then trivial, since we have $\int \theta_\lambda = 0$ in those cases. For the other applications we mentioned, additional work is necessary.

To conclude, let us say that, provided the class $\mathcal{C}$ is slightly extended, every Calderón-Zygmund operator can be decomposed into a sum $T_1 + T_2^*$, where $T_1, T_2 \in \mathcal{C}$. This decomposition is the key point in the celebrated $T(1)$-theorem ([**6**]), which characterizes the L^2-continuity of operators whose kernels are of Calderón-Zygmund type by the three properties:

    i) T has the Weak Boundedness Property (in terms of wavelets, it means the existence of $C > 0$ such that $|\langle T\varphi_{jk}, \ \varphi_{j\ell}\rangle| \leq C$ for every j, k, ℓ),

    ii) $T(1) \in BMO$,

    iii) $T^*(1) \in BMO$.

$(T(1)$ and $T^*(1)$ are defined as functionals on a suitable test space.)

In the special case where T belongs to $\mathcal{C}$, one has $T(1) = 0$ and

$$T^*(1) = \sum_\lambda \left(\overline{\int \theta_\lambda} \right) \psi_\lambda \,,$$

so that Carleson's condition on $\int \theta_\lambda$ is nothing but the characterization of BMO with wavelet coefficients.

REFERENCES

1. P. Auscher, Ph. Tchamitchian, "Ondelettes et conjecture de Kato", CRAS Vol. 313, Série I (1991) pp. 63-66.

2. P. Auscher, Ph. Tchamitchian, "Conjecture de Kato sur les ouverts de $\mathbb{R}$", *Rev. Mat. Iberoamericana*, Vol. 8, No. 2, (1992) pp. 149-199.

3. C. Basdevant, V. Perrier, "La décomposition en ondelettes périodiques, un outil pour l'analyse des champs inhomogènes", *Théorie et algorithmes. La Recherche Aérospatiale* **3** (1989) pp. 53-67.

4. A. P. Calderón, "Cauchy problem for partial differential equations", *Amer. J. Math.* **80** (1958).

5. R. R. Coifman and Y. Meyer, "Au-delà des opératerus pseudo-differéntiees", *Astérisque* **57** (1978).

6. G. David and J. L. Journé, "A boundedness criterion for generalized Calderón-Zygmund operators", *Ann. Math.* **120** (1984) pp. 371–397.

7. S. Jaffard, "Propriétés des matrices 'bien localisées' près de leur diagonale et quelques applications", *Ann. Inst. H. Poincaré, Analyse nor linéaire* **7** (1990), pp. 461-476.

8. S. Lazaar, J. Liandrat, Ph. Tchamitchian, "Détermination itérative de l'inverse d'opérateurs elliptiques à coefficients variables", submitted to *C. R. Acad. Sci. Paris*, Série 1, 1993.

9. Y. Meyer, "Ondelettes et opérateurs", I, II, III, Hermann (Paris), 1990.

10. Ph. Tchamitchian, "Bases d'ondelettes et intégrales singulierès: analyse des functions et calcul sur les opérateurs", Habilitation à diriger des recherches, Université Aix-Marseille II, Faculté des Sciences de Luminy, 1989.

MATHÉMATIQUES, FACULTÉ DES SCIENCES, UNIVERSITÉ DE MARSEILLE À SAINT-JÉROME, CASE 322, AVENUE ESCADRILLE NORMANDIE NIEMEN, 13397 MARSEILLE CÉDEX 20, FRANCE

E-mail: tchamphi@frmrs11.bitnet

Proceedings of Symposia in Applied Mathematics
Volume **47**, 1993

Wavelets and Fast Numerical Algorithms

GREGORY BEYLKIN

ABSTRACT. Wavelet-based algorithms in numerical analysis are similar
to other transform methods in that vectors and operators are expanded
into a basis and the computations take place in the new system of coordinates. However, wavelet-based algorithms exhibit a number of important
properties due to the controllable localization of wavelets in both time and
frequency domains and their orthogonality to low degree polynomials. The
multiresolution structure of the wavelet expansions brings about an efficient organization of transformations. Moreover, wide classes of operators
(Calderón-Zygmund operators, for example) which naively would require a
full (dense) matrix for their numerical description, have sparse representations in wavelet bases. For these operators sparse representations lead to
fast numerical algorithms, e.g. an $O(-N \log \varepsilon)$ algorithm for the evaluation
of $N \times N$ matrices on vectors, or an $O(-N \log \varepsilon)$ algorithm for matrix multiplications, where ε is the desired accuracy. Since the performance of many
algorithms requiring multiplication of dense matrices has been limited by
$O(N^3)$ operations, these fast algorithms address a critical numerical issue.
In this lecture, we review the standard and non-standard representations of
operators in wavelet bases and associated fast numerical algorithms. The
non-standard representation uncouples the interaction among the scales.
Examples of the non-standard forms of several basic operators are computed explicitly.

Numerical algorithms using wavelet bases are similar to other transform methods in that vectors and operators are expanded into a basis and the computations
take place in the new system of coordinates. As with all transform methods, such
an approach hopes to achieve that the computation is faster in the new system of
coordinates than in the original domain. However, due to the recursive definition
of wavelets, their controllable localization in both space and wave number (time
and frequency) domains, and the vanishing moments property, wavelet based
algorithms exhibit a number of new and important properties.

1991 *Mathematics Subject Classification.* Primary 65R20, 42C15, 45L10, 65D99.

The research was partially supported by ONR grant N00014-91-J4037 and a grant from
Chevron Oil Field Research Company.

© 1993 American Mathematical Society
0160-7634/93 $1.00 + $.25 per page

In the usual transform methods, the functions of the basis (e.g. exponentials, Chebyshev polynomials, etc.) are chosen to be eigenfunctions of some differential operator (e.g. solutions of the Sturm-Liouville problem). The choice of the differential operator and, hence, of the basis functions, is dictated by the availability of fast algorithms for expanding an arbitrary function into the basis. Unfortunately, classes of operators which have a sparse representation in such bases are very narrow.

Wavelets, on the other hand, are not solutions of a differential equation. These functions are defined recursively and are generated via an iterative algorithm. They are translations and dilations of a single function.[1] Instead of diagonalizing some differential operator, representations in the wavelet bases reduce a wide class of operators to a sparse form. Here the orthogonality of wavelets to the low degree polynomials (the vanishing moments property) plays a crucial role in producing sparse systems.[2]

Historically, the orthonormal bases of wavelets were first constructed by Stromberg [33] and then by Meyer [25]. Later, the notion of the Multiresolution Analysis was introduced by Meyer [26] and Mallat [23]. Orthonormal bases of compactly supported wavelets were constructed by Daubechies [16]. There are many new constructions of orthonormal bases with a controllable localization in the time–frequency domain, notably "wavelet-packet" bases in [13] and [15], local trigonometric bases in [14] and [24], and wavelet bases on the interval in [11], [12] and [22]. There exists a very important connection between wavelets and the technique of subband coding in signal processing. In fact, the discrete wavelet transform is accomplished by a pair of so-called quadrature mirror filters. Quadrature mirror filters (QMFs) with the "exact reproduction property" were introduced by Smith and Barnwell [32].

Wavelets have some of their historical roots in Littlewood-Paley and Calderón-Zygmund theories (see e.g. [28]) which have been powerful tools in the analysis of linear and non-linear operators. In numerical analysis some of the ingredients of Calderón-Zygmund theory appear in the Fast Multipole Method (FMM) for computing potential interactions [30], [19], [10]. FMM was designed for computing potential interactions between N particles in $O(-N \log \varepsilon)$ operations (instead of $O(N^2)$ operations). The reduction of the complexity in FMM is achieved by approximating the far field effect of a cloud of charges located in a box by the effect of a single multipole at the center of the box. All boxes are then organized in a dyadic hierarchy enabling an efficient $O(N)$ algorithm.

The fast wavelet-based algorithms of [7] provide a systematic generalization of the FMM and its descendents (e.g. [29], [2], [18]) to all Calderón-Zygmund and pseudo-differential operators. The subdivision of the space and its organization

---

[1]It is also possible to construct bases with translations and dilations of several functions, see e.g. [1].

[2]This property and the fact that the basis is orthonormal, distinguish the wavelet bases from the hierarchical bases.

in a dyadic hierarchy are a consequence of the multiresolution properties of the wavelet bases, while the vanishing moments of the basis functions make them useful tools for approximation.

A novel aspect of representing operators in the wavelet bases is the so-called non-standard form [7]. The remarkable feature of the non-standard form is the uncoupling of the interactions between the scales. The non-standard form leads to an order N algorithm for evaluating operators on functions. It is also quite remarkable that the error estimates for the non-standard form lead to a proof of the celebrated "T(1)" theorem of David and Journé (see [7]). The non-standard forms of many basic operators, such as derivatives, fractional derivatives, the Hilbert and Riesz transforms, may be computed explicitly [4]. A straightforward realization, or the standard form, by contrast, contains matrix entries reflecting "interactions" between all pairs of scales. The standard form yields, in general, only an order $N \log(N)$ algorithm for evaluating operators on functions.

The representation of wide classes of operators in wavelet bases may be viewed as a method for their "compression", i.e., conversion to a sparse form. For these operators sparse representations lead to fast algorithms for matrix multiplications. Since the performance of many algorithms requiring multiplication of dense matrices has been limited by $O(N^3)$ operations, these fast algorithms address a critical numerical issue.

Examples of such algorithms requiring multiplication of matrices are, for instance, an iterative algorithm for constructing the generalized inverse [31], the scaling and squaring method for computing the exponential of an operator, and similar algorithms for sine and cosine of an operator, to mention a few. By replacing the ordinary matrix multiplication in these algorithms by the fast multiplication in the wavelet bases, the number of operations is reduced to, essentially, $O(N)$ operations. For example, if both the operator and its generalized inverse admit sparse representations in the wavelet basis, then the iterative algorithm [31] for computing the generalized inverse requires only $O(N \log \kappa)$ operations, where κ is the condition number of the matrix. Various numerical examples and applications may be found in [9], [1] and [8].

Solving the two-point boundary value problem for the elliptic differential operators in the wavelet "system of coordinates" allows us to construct the Green's function (the inverse operator) in $O(N)$ operation. We note that the ordinary matrix representation of the Green's function requires $O(N^2)$ significant entries but the representation of the Green's function in the wavelet bases requires (for a given accuracy) only $O(N)$ entries. The main tool in constructing the Green's function numerically is the diagonal preconditioner available for periodized differential operators in the wavelet bases [4], [5] (see also [21]).

Unfortunately, the format of one lecture does not allow us to cover all the developments or mention all the results available today. Instead, we will review several features of the new numerical methodology based on the wavelet representations. Starting from the notion of multiresolution analysis, we will consider the non-standard form (which achieves uncoupling among the scales) and the associated fast numerical algorithms. Examples of non-standard forms of several basic operators (e.g. derivatives) will be computed explicitly.

1. Multiresolution analysis and wavelets.

We briefly outline here the properties of compactly supported wavelets and refer for details to [16], [17] and [28]. Let us start with the notion of a multiresolution analysis [26], [23] which captures the essential features of all multiresolution approaches developed so far.

DEFINITION 1.1. A multiresolution analysis is a decomposition of the Hilbert space $\mathbf{L}^2(\mathbf{R^d})$, $\mathbf{d} \geq 1$, into a chain of closed subspaces

$$(1.1) \qquad \cdots \subset \mathbf{V}_2 \subset \mathbf{V}_1 \subset \mathbf{V}_0 \subset \mathbf{V}_{-1} \subset \mathbf{V}_{-2} \subset \cdots$$

such that
  (i) $\bigcap_{j \in \mathbf{Z}} \mathbf{V}_j = \{0\}$ and $\bigcup_{j \in \mathbf{Z}} \mathbf{V}_j$ is dense in $\mathbf{L}^2(\mathbf{R^d})$.
  (ii) For any $f \in \mathbf{L}^2(\mathbf{R^d})$ and any $j \in \mathbf{Z}$, $f(x) \in \mathbf{V}_j$ if and only if $f(2x) \in \mathbf{V}_{j-1}$.
  (iii) For any $f \in \mathbf{L}^2(\mathbf{R^d})$ and any $k \in \mathbf{Z^d}$, $f(x) \in \mathbf{V}_0$ if and only if $f(x-k) \in \mathbf{V}_0$.
  (iv) There exists a function $\varphi \in \mathbf{V}_0$ such that $\{\varphi(x-k)\}_{k \in \mathbf{Z^d}}$ is a Riesz basis of $\mathbf{V}_0$.

In this lecture we use only orthonormal bases, so that we replace Condition iv by

  (iv') There exists a function $\varphi \in \mathbf{V}_0$ such that $\{\varphi(x-k)\}_{k \in \mathbf{Z^d}}$ is an orthonormal basis of $\mathbf{V}_0$.

Let us define the subspaces $\mathbf{W}_j$ as an orthogonal complement of $\mathbf{V}_j$ in $\mathbf{V}_{j-1}$,

$$(1.2) \qquad \mathbf{V}_{j-1} = \mathbf{V}_j \oplus \mathbf{W}_j,$$

and represent the space $\mathbf{L}^2(\mathbf{R^d})$ as a direct sum

$$(1.3) \qquad \mathbf{L}^2(\mathbf{R^d}) = \bigoplus_{j \in \mathbf{Z}} \mathbf{W}_j.$$

Selecting the coarsest scale n, we may replace the chain of the subspaces (1.1) by

$$(1.4) \qquad \mathbf{V}_n \subset \cdots \subset \mathbf{V}_2 \subset \mathbf{V}_1 \subset \mathbf{V}_0 \subset \mathbf{V}_{-1} \subset \mathbf{V}_{-2} \subset \ldots,$$

and obtain

$$(1.5) \qquad \mathbf{L}^2(\mathbf{R^d}) = \mathbf{V}_n \bigoplus_{j \leq n} \mathbf{W}_j.$$

If there is a finite number of scales then, without loss of generality, we set $j = 0$ to be the finest scale and consider

(1.6) $$\mathbf{V}_n \subset \cdots \subset \mathbf{V}_2 \subset \mathbf{V}_1 \subset \mathbf{V}_0, \quad \mathbf{V}_0 \subset \mathbf{L}^2(\mathbf{R^d})$$

instead of (1.4). In numerical realizations the subspace $\mathbf{V}_0$ is finite dimensional.

First, let us consider bases in $L^2(\mathbb{R})$, $d = 1$. The function φ is the so-called *scaling function* and, with its help, we may define the function ψ, the *wavelet*, such that the set of functions $\{\psi(x - k)\}_{k \in \mathbf{Z}}$ is an orthonormal basis of $\mathbf{W}_0$.

An example of a multiresolution analysis satisfying Definition 1.1 with Condition iv' is the chain of subspaces generated by the Haar basis [20]. The scaling function in this case is the characteristic function of the interval $(0, 1)$. The Haar function is defined as

(1.7) $$h(x) = \left\{ \begin{array}{ll} 1 & \text{for } 0 < x < 1/2 \\ -1 & \text{for } 1/2 \leq x < 1 \\ 0 & \text{elsewhere.} \end{array} \right.$$

and the Haar basis is formed by functions $h_{j,k}(x) = 2^{-j/2}h(2^{-j}x - k)$, $j, k \in \mathbf{Z}$.

Wavelet bases (with a smooth scaling function φ in Condition iv') generalizing the Haar functions were first constructed by Stromberg [33] and then Meyer [25]. The notion of the Multiresolution Analysis was introduced by Meyer [26] and Mallat [23] and is more recent than the constructions of [33], [25] and, of course, of [20]. Compactly supported wavelets with vanishing moments were constructed by I. Daubechies [16]; those are the ones we will use in this lecture. However, most of the results that we discuss do not depend on this particular choice of the wavelet basis.

The vanishing moments property simply means that the basis functions are chosen to be orthogonal to the low degree polynomials, namely, if the set of functions $\{\psi(x - k)\}_{k \in \mathbf{Z}}$ is an orthonormal basis of $\mathbf{W}_0$, then

(1.8) $$\int_{-\infty}^{+\infty} \psi(x)x^m dx = 0, \qquad m = 0, \ldots, M - 1.$$

For the Haar function in (1.7) $M = 1$ and it is indeed trivially orthogonal to constants.

There are two immediate consequences of Definition 1.1 with Condition iv'. First, the function φ may be expressed as a linear combination of the basis functions of $\mathbf{V}_{-1}$. Since the functions $\{\varphi_{j,k}(x) = 2^{-j/2}\varphi(2^{-j}x - k)\}_{k \in \mathbf{Z}}$ form an orthonormal basis of $\mathbf{V}_j$, we have

(1.9) $$\varphi(x) = \sqrt{2} \sum_{k=0}^{L-1} h_k \varphi(2x - k).$$

In general, the sum in (1.9) does not have to be finite and, by choosing a finite sum in (1.9), we are selecting compactly supported wavelets. We may rewrite

(1.9) as

(1.10) $$\hat{\varphi}(\xi) = m_0(\xi/2)\hat{\varphi}(\xi/2),$$

where

(1.11) $$\hat{\varphi}(\xi) = \frac{1}{\sqrt{2\pi}} \int_{-\infty}^{+\infty} \varphi(x)\, e^{ix\xi}\, dx,$$

and the 2π-periodic function m_0 is defined as

(1.12) $$m_0(\xi) = 2^{-1/2} \sum_{k=0}^{L-1} h_k e^{ik\xi}.$$

Second, the orthogonality of $\{\varphi(x-k)\}_{k\in\mathbf{Z}}$ implies that

(1.13) $$\delta_{k0} = \int_{-\infty}^{+\infty} \varphi(x-k)\varphi(x)\, dx = \int_{-\infty}^{+\infty} |\hat{\varphi}(\xi)|^2\, e^{-ik\xi}\, d\xi,$$

and, therefore,

(1.14) $$\delta_{k0} = \int_0^{2\pi} \sum_{l\in\mathbf{Z}} |\hat{\varphi}(\xi+2\pi l)|^2\, e^{-ik\xi}\, d\xi,$$

and

(1.15) $$\sum_{l\in\mathbf{Z}} |\hat{\varphi}(\xi+2\pi l)|^2 = \frac{1}{2\pi}.$$

Using (1.10), we obtain

(1.16) $$\sum_{l\in\mathbf{Z}} |m_0(\xi/2+\pi l)|^2 |\hat{\varphi}(\xi/2+\pi l)|^2 = \frac{1}{2\pi},$$

and, by taking the sum in (1.16) separately over odd and even indices, we have

(1.17) $$\sum_{l\in\mathbf{Z}} |m_0(\xi/2+2\pi l)|^2 |\hat{\varphi}(\xi/2+2\pi l)|^2$$

$$+ \sum_{l\in\mathbf{Z}} |m_0(\xi/2+2\pi l+\pi)|^2 |\hat{\varphi}(\xi/2+2\pi l+\pi)|^2 = \frac{1}{2\pi}.$$

Using the 2π-periodicity of the function m_0 and (1.15), we obtain (after replacing $\xi/2$ by ξ) a necessary condition

(1.18) $$|m_0(\xi)|^2 + |m_0(\xi+\pi)|^2 = 1,$$

for the coefficients h_k in (1.12). On defining the function ψ by

(1.19) $$\psi(x) = \sqrt{2} \sum_k g_k \varphi(2x-k),$$

where

(1.20) $$g_k = (-1)^k h_{L-k-1}, \qquad k = 0, \ldots, L-1,$$

or, equivalently, the Fourier transform of ψ by

$$(1.21) \qquad \hat{\psi}(\xi) = m_1(\xi/2)\hat{\varphi}(\xi/2),$$

where

$$(1.22) \qquad m_1(\xi) = 2^{-1/2} \sum_{k=0}^{k=L-1} g_k e^{ik\xi} = e^{-i\xi}\overline{m}_0(\xi + \pi),$$

it is not difficult to show (see e.g., [28], [16], [17]), that for each fixed scale $j \in \mathbf{Z}$, the wavelets $\{\psi_{j,k}(x) = 2^{-j/2}\psi(2^{-j}x - k)\}_{k\in\mathbf{Z}}$ form an orthonormal basis of $\mathbf{W}_j$.

Equation (1.18) can also be viewed as the condition for exact reconstruction for a pair of the quadrature mirror filters (QMFs) H and G, where $H = \{h_k\}_{k=0}^{k=L-1}$ and $G = \{g_k\}_{k=0}^{k=L-1}$. Such exact QMF filters were first introduced by Smith and Barnwell [32] for subband coding.

We will not go into a full discussion of the necessary and sufficient conditions for the quadrature mirror filters H and G to generate a wavelet basis and refer to [17] for the details. The coefficients of the quadrature mirror filters H and G are computed by solving a set of algebraic equations (see e.g. [17]). The number L of the filter coefficients in (1.12) and (1.22) is related to the number of vanishing moments M, and $L = 2M$ for the wavelets constructed in [16]. If additional conditions are imposed (see [7] for an example), then the relation might be different, but L is always even.

We observe that once the filter H has been chosen, it completely determines the functions φ and ψ and therefore, the multiresolution analysis. Moreover, in properly constructed algorithms, the values of the functions φ and ψ are (almost) never computed. Due to the recursive definition of the wavelet bases, all the manipulations are performed with the quadrature mirror filters H and G, even if they involve quantities associated with φ and ψ.

As an example, let us compute the moments of the scaling function ϕ. The expressions for the moments,

$$(1.23) \qquad \mathcal{M}_m = \int x^m\,\varphi(x)\,dx, \quad m = 0,\ldots,M-1,$$

may be found in terms of the filter coefficients $\{h_k\}_{k=0}^{k=L-1}$. Applying the operator $(\frac{1}{i}d/d\xi)^m$ to both sides of (1.10) and setting $\xi = 0$, we obtain

$$(1.24) \qquad \mathcal{M}_m = 2^{-m} \sum_{j=0}^{j=m} \binom{m}{j} \mathcal{M}_j \mathcal{M}_{m-j}^h,$$

where

$$(1.25) \qquad \mathcal{M}_l^h = 2^{-\frac{1}{2}} \sum_{k=0}^{k=L-1} h_k k^l, \quad l = 0,\ldots,M-1.$$

Thus, we have from (1.24)

$$(1.26) \qquad \mathcal{M}_m = \frac{1}{2^m - 1} \sum_{j=0}^{j=m-1} \binom{m}{j} \mathcal{M}_j \mathcal{M}_{m-j}^h,$$

with $\mathcal{M}_0 = 1$.

Alternatively, using

$$(1.27) \qquad \hat{\varphi}(\xi) = (2\pi)^{-1/2} \prod_{j=1}^{\infty} m_0(2^{-j}\xi),$$

the moments $\mathcal{M}_m$ may be obtained within the desired accuracy as a limit of a recursively generated sequence of vectors, $\{\mathcal{M}_m^{(r)}\}_{m=0}^{m=M-1}$ for $r = 1, 2, \ldots$,

$$(1.28) \qquad \mathcal{M}_m^{(r+1)} = \sum_{j=0}^{j=m} \binom{m}{j} 2^{-j(r+1)} \mathcal{M}_{m-j}^{(r)} \mathcal{M}_j^h,$$

starting with

$$(1.29) \qquad \mathcal{M}_m^{(1)} = 2^{-m} \mathcal{M}_m^h, \quad m = 0, \ldots, M - 1.$$

Each vector $\{\mathcal{M}_m^{(r)}\}_{m=0}^{m=M-1}$ represents M moments of the product in (1.27) with r terms, and the iteration converges rapidly. Notice that in both algorithms we never computed the values of the function φ itself.

2. The non-standard form.

A wavelet basis in $\mathbf{L}^2(\mathbf{R}^\mathbf{d})$, $d \geq 2$, may be constructed as a tensor product of one-dimensional bases. Considering $d = 2$ and using the Haar basis as an example, we note that the supports of the basis functions are rectangles of various dyadic sizes. Representing operators in such bases leads to the standard form which we will discuss in the next section.

Alternatively, wavelet bases in $\mathbf{L}^2(\mathbf{R}^\mathbf{d})$, $d \geq 2$ may be constructed using the scaling function in addition to the wavelets. Such a construction is specific to wavelet bases. Considering $d = 2$ as an example, we note that the triplet of functions

$$(2.1) \qquad \{\psi_{j,k}(x)\,\psi_{j,k'}(y), \ \psi_{j,k}(x)\,\varphi_{j,k'}(y), \ \varphi_{j,k}(x)\,\psi_{j,k'}(y)\},$$

where $j, k, k' \in \mathbf{Z}$, forms a basis of $\mathbf{L}^2(\mathbf{R}^\mathbf{2})$. We note that the basis functions have square supports. Representing operators in these bases leads to the non-standard form [7].

Let us introduce the non-standard form in the context of Multiresolution Analysis, independently of the specific choice of the wavelet basis. Let T be an operator

$$(2.2) \qquad T : \mathbf{L}^2(\mathbf{R}) \to \mathbf{L}^2(\mathbf{R}),$$

with kernel $K(x,y)$. The orthogonal projection operators on the subspace $\mathbf{V}_j$, $j \in \mathbf{Z}$,

$$(2.3) \qquad P_j : \mathbf{L}^2(\mathbf{R}) \to \mathbf{V}_j,$$

are given by

$$(2.4) \qquad (P_j f)(x) = \sum_k \langle f, \varphi_{j,k} \rangle \varphi_{j,k}(x).$$

Expanding T in a "telescopic" series, we obtain

$$(2.5) \qquad T = \sum_{j \in \mathbf{Z}} (Q_j T Q_j + Q_j T P_j + P_j T Q_j),$$

where

$$(2.6) \qquad Q_j = P_{j-1} - P_j$$

is the projection operator on the subspace $\mathbf{W}_j$. If there is a coarsest scale n, then instead of (2.5) we have

$$(2.7) \qquad T = \sum_{j=-\infty}^{n} (Q_j T Q_j + Q_j T P_j + P_j T Q_j) + P_n T P_n,$$

and if the scale $j = 0$ is the finest scale, then

$$(2.8) \qquad T_0 = \sum_{j=1}^{n} (Q_j T Q_j + Q_j T P_j + P_j T Q_j) + P_n T P_n,$$

where $T \sim T_0 = P_0 T P_0$ is a discretization of the operator T on the finest scale.

The non-standard form is a representation (see [**7**]) of the operator T as a chain of triplets

$$(2.9) \qquad T = \{A_j, B_j, \Gamma_j\}_{j \in \mathbf{Z}}$$

acting on the subspaces $\mathbf{V}_j$ and $\mathbf{W}_j$,

$$(2.10) \qquad A_j : \mathbf{W}_j \to \mathbf{W}_j,$$

$$(2.11) \qquad B_j : \mathbf{V}_j \to \mathbf{W}_j,$$

$$(2.12) \qquad \Gamma_j : \mathbf{W}_j \to \mathbf{V}_j.$$

The operators $\{A_j, B_j, \Gamma_j\}_{j \in \mathbf{Z}}$ are defined as $A_j = Q_j T Q_j$, $B_j = Q_j T P_j$ and $\Gamma_j = P_j T Q_j$. These operators admit a recursive definition via the relation

$$(2.13) \qquad T_j = \begin{pmatrix} A_{j+1} & B_{j+1} \\ \Gamma_{j+1} & T_{j+1} \end{pmatrix},$$

where the operators T_j,

$$(2.14) \qquad\qquad T_j : \mathbf{V}_j \to \mathbf{V}_j,$$

are defined by $T_j = P_j T P_j$.

  If there is a coarsest scale n, then

$$(2.15) \qquad\qquad T = \{\{A_j, B_j, \Gamma_j\}_{j \in \mathbf{Z}: j \leq n}, T_n\},$$

where $T_n = P_n T P_n$. If the number of scales is finite, then $j = 1, 2, \ldots, n$ in (2.15) and the operators are organized as blocks of a matrix (see Figures 1 and 2).

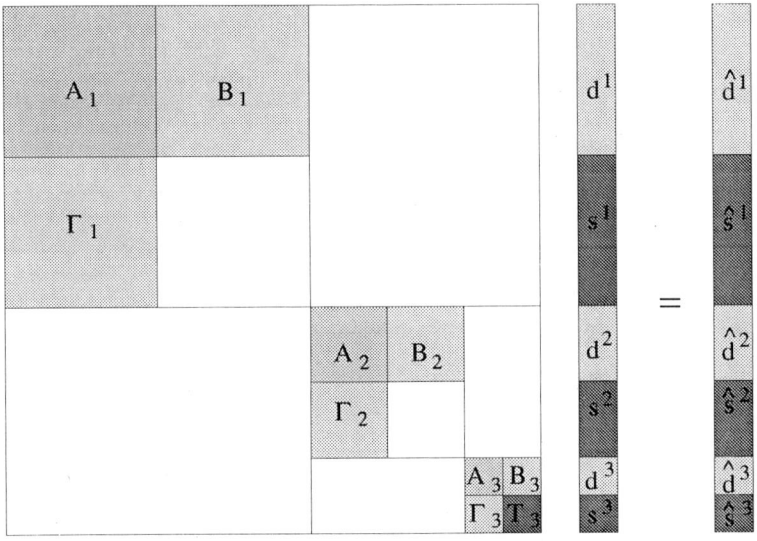

FIGURE 1. Organization of the non-standard form of a matrix. The submatrices A_j, B_j, and Γ_j, $j = 1, 2, 3$, and T_3 are the only non-zero submatrices.

Let us make the following observations:

(i) The map (2.10) implies that the operator A_j describes the interaction on the scale j only, since the subspace $\mathbf{W}_j$ is an element of the direct sum in (1.5).

(ii) The operators B_j, Γ_j in (2.11) and (2.12) describe the interaction between scale j and all coarser scales. Indeed, the subspace $\mathbf{V}_j$ contains all the subspaces $\mathbf{V}_{j'}$ with $j' > j$ (see (1.1)).

(iii) The operator T_j is an "averaged" version of the operator T_{j-1}.

  The operators A_j, B_j and Γ_j are represented by the matrices α^j, β^j and γ^j,

$$(2.16) \qquad\qquad \alpha^j_{k,k'} = \int \int K(x, y)\, \psi_{j,k}(x)\, \psi_{j,k'}(y)\, dx dy,$$

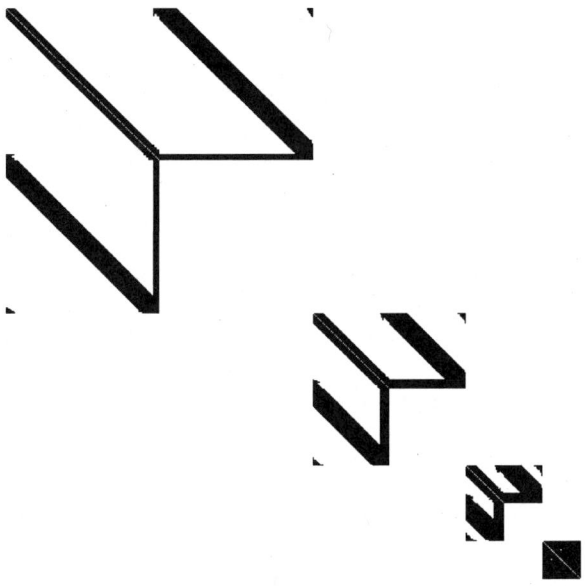

FIGURE 2. An example of a matrix in the non-standard form (see Example 4.2).

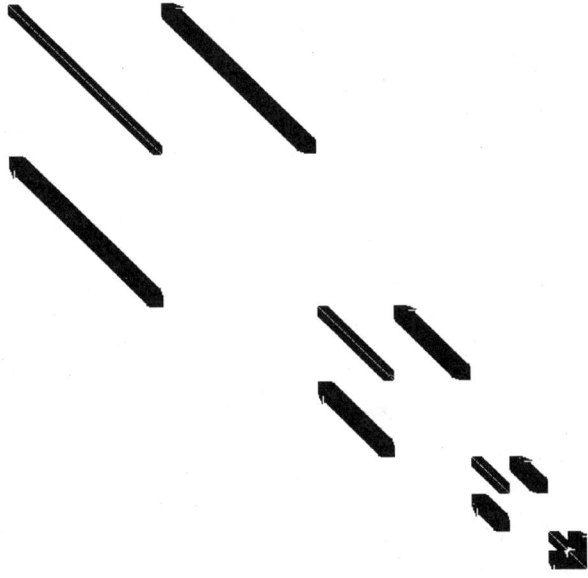

FIGURE 3. The non-standard form of the same matrix as in Figure 2 using a basis of wavelets on the interval [**6**]. The vertical and horizontal bands (which are present in Figure 2 due to periodization) do not appear in this representation.

$$(2.17) \qquad \beta_{k,k'}^{j} = \int \int K(x,y)\,\psi_{j,k}(x)\,\varphi_{j,k'}(y)\,dxdy,$$

and

$$(2.18) \qquad \gamma_{k,k'}^{j} = \int \int K(x,y)\,\varphi_{j,k}(x)\,\psi_{j,k'}(y)\,dxdy.$$

The operator T_j is represented by the matrix s^j,

$$(2.19) \qquad s_{k,k'}^{j} = \int \int K(x,y)\,\varphi_{j,k}(x)\,\varphi_{j,k'}(y)\,dxdy.$$

3. The standard form.

The standard form is the representation of an operator in the tensor product basis. Instead of introducing the standard form in this manner, we emphasize the connection with the non-standard form. The standard form is obtained by representing

$$(3.1) \qquad \mathbf{V}_j = \bigoplus_{j'>j} \mathbf{W}_{j'},$$

and considering for each scale j the operators $\{B_j^{j'}, \Gamma_j^{j'}\}_{j'>j}$,

$$(3.2) \qquad B_j^{j'} : \mathbf{W}_{j'} \to \mathbf{W}_j,$$

$$(3.3) \qquad \Gamma_j^{j'} : \mathbf{W}_j \to \mathbf{W}_{j'}.$$

If there is a coarsest scale n, then instead of (3.1) we have

$$(3.4) \qquad \mathbf{V}_j = \mathbf{V}_n \bigoplus_{j'=j+1}^{j'=n} \mathbf{W}_{j'}.$$

In this case, the operators $\{B_j^{j'}, \Gamma_j^{j'}\}$ for $j' = j+1, \dots, n$ are as in (3.2) and (3.3) and, in addition, for each scale j there are operators $\{B_j^{n+1}\}$ and $\{\Gamma_j^{n+1}\}$,

$$(3.5) \qquad B_j^{n+1} : \mathbf{V}_n \to \mathbf{W}_j,$$

$$(3.6) \qquad \Gamma_j^{n+1} : \mathbf{W}_j \to \mathbf{V}_n.$$

(In this notation, $\Gamma_n^{n+1} = \Gamma_n$ and $B_n^{n+1} = B_n$). If there are finitely many scales and $\mathbf{V}_0$ is finite dimensional, then the standard form is a representation of $T_0 = P_0 T P_0$ as

$$(3.7) \qquad T_0 = \{A_j, \{B_j^{j'}\}_{j'=j+1}^{j'=n}, \{\Gamma_j^{j'}\}_{j'=j+1}^{j'=n}, B_j^{n+1}, \Gamma_j^{n+1}, T_n\}_{j=1,\dots,n}.$$

The operators (3.7) can again be organized as blocks of a matrix (see Figures 4 and 5).

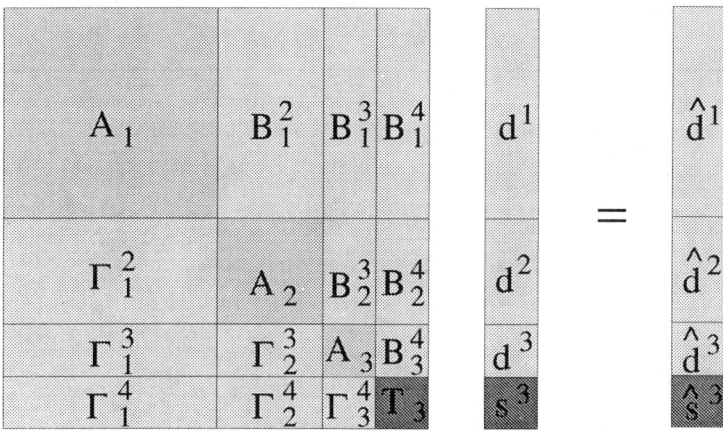

FIGURE 4. Organization of a matrix in the standard form.

If the operator T is a Calderón-Zygmund or a pseudo-differential operator then, for a fixed accuracy, all the operators in (3.7) (except T_n) are banded. As a result, the standard form has several "finger" bands which correspond to the interaction between different scales. For a large class of operators (pseudo-differential, for example), the interaction between different scales, characterized by the size of the coefficients of "finger" bands, decays as the distance $j' - j$ between the scales increases. Therefore, if the scales j and j' are well separated, then for a given accuracy, the operators $B_j^{j'}, \Gamma_j^{j'}$ can be neglected.

There are two ways of computing the standard form of a matrix. The first consists in applying the one-dimensional transform to each column (row) of the matrix and, then, to each row (column) of the result. Alternatively, one can compute the non-standard form and then apply the one-dimensional transform to each row of all operators B^j and each column of all operators Γ_j. We refer to [**7**] for details.

4. Compression of operators.

If the operator T is a Calderón-Zygmund or a pseudo-differential operator then, by using the wavelet basis with M vanishing moments, we force the entries of the matrices $\{A_j, B_j, \Gamma_j\}_{j \in \mathbf{Z}}$ to decay roughly as $1/d^{M+1}$, where d is the distance from the diagonal. For example, let the kernel satisfy the conditions

$$(4.1) \qquad\qquad |K(x,y)| \;\leq\; \frac{1}{|x-y|},$$

$$(4.2) \qquad |\partial_x^M K(x,y)| + |\partial_y^M K(x,y)| \;\leq\; \frac{C_0}{|x-y|^{1+M}}$$

FIGURE 5. An example of a matrix in the standard form (see Example 4.2).

for some $M \geq 1$. Then by choosing the wavelet basis with M vanishing moments, the coefficients $\alpha_{i,l}^j, \beta_{i,l}^j, \gamma_{i,l}^j$ of the non-standard form (see (2.16) - (2.18)) satisfy the estimate

$$(4.3) \qquad |\alpha_{i,l}^j| + |\beta_{i,l}^j| + |\gamma_{i,l}^j| \leq \frac{C_M}{1 + |i - l|^{M+1}},$$

for all

$$(4.4) \qquad |i - l| \geq 2M.$$

If, in addition to (4.1), (4.2),

$$(4.5) \qquad \left| \iint_{I \times I} K(x,y) \, dx dy \right| \leq C|I|$$

for all dyadic intervals I (this is the "weak cancellation condition", see [27]), then (4.3) holds for all i, l.

If T is a pseudo-differential operator with symbol $\sigma(x, \xi)$ of order λ defined by the formula

$$(4.6) \quad T(f)(x) = \sigma(x, D)f = \int e^{ix\xi} \, \sigma(x, \xi) \hat{f}(\xi) \, d\xi = \int K(x,y) f(y) \, dy,$$

where K is the distributional kernel of T, then assuming that the symbols σ of T and σ^* of T^* satisfy the standard conditions

$$(4.7) \qquad |\partial_\xi^\alpha \partial_x^\beta \sigma(x,\xi)| \leq C_{\alpha,\beta}(1+|\xi|)^{\lambda-\alpha+\beta}$$

$$(4.8) \qquad |\partial_\xi^\alpha \partial_x^\beta \sigma^*(x,\xi)| \leq C_{\alpha,\beta}(1+|\xi|)^{\lambda-\alpha+\beta},$$

we have the inequality

$$(4.9) \qquad |\alpha_{i,l}^j| + |\beta_{i,l}^j| + |\gamma_{i,l}^j| \leq \frac{2^{\lambda j} C_M}{(1+|i-l|)^{M+1}},$$

for all integer i,l.

Suppose now that we approximate the operator T_0 by the operator T_0^B obtained from T_0 by setting to zero all coefficients of matrices α^j, β^j and γ^j outside bands of width $B \geq 2M$ around their diagonals. We obtain

$$(4.10) \qquad \| T_0^B - T_0 \| \leq \frac{C}{B^M} \log_2 N,$$

where C is a constant determined by the kernel K and $\log_2 N$ is the number of scales in the representation. In most numerical applications, the accuracy ε of calculations is fixed, and the parameters of the algorithm (in our case, the band width B and the order M) have to be chosen in such a manner that the desired computational precision is achieved. If M is fixed, then we choose B so that

$$(4.11) \qquad B \geq \left(\frac{C}{\varepsilon} \log_2 N\right)^{1/M}.$$

In other words, T_0 has been approximated to precision ε with its truncated version, which can be applied to arbitrary vectors for a cost proportional to $N\left((C/\varepsilon) \log_2 N\right)^{1/M}$, which for all practical purposes does not differ from N.

A more detailed investigation [7] permits the estimate (4.10) to be replaced with the estimate

$$(4.12) \qquad \| T_0^B - T_0 \| \leq \frac{C}{B^M},$$

making the application of the operator T_0 to an arbitrary vector with arbitrary fixed accuracy into a procedure of order N. Obtaining this uniform estimate leads to a proof of

THEOREM 4.1 (G. DAVID, J.L. JOURNÉ). Suppose that the operator

$$(4.13) \qquad T(f) = \int K(x,y) f(y)\, dy$$

satisfies the conditions (4.1), (4.2), (4.5). Then a necessary and sufficient condition for T to be bounded on L^2 is that

$$(4.14) \qquad \beta(x) = T(1)(x),$$

(4.15) $\gamma(y) = T^*(1)(y)$

satisfy the dyadic bounded mean oscillation (B.M.O.) condition,

(4.16) $\displaystyle \sup_J \frac{1}{|J|} \int_J |\beta(x) - m_J(\beta)|^2 dx \leq C,$

where J is a dyadic interval and

(4.17) $\displaystyle m_J(\beta) = \frac{1}{|J|} \int_J \beta(x) dx.$

Again we refer to [**7**] for details.

The compression of operators results in fast algorithms for evaluation of operators on functions. We present here one example and refer to [**7**] for additional examples.

EXAMPLE 4.2. In this example, we consider the matrix

$$A_{ij} = \begin{cases} \frac{1}{i-j} & i \neq j, \\ 0 & i = j, \end{cases}$$

and convert it to the non-standard form using wavelets with six vanishing moments. Setting to zero all entries whose absolute values are smaller than 10^{-7}, we obtain the non-standard form where the non-zero elements are shown in black in Figure 2. The results of experiments in applying this sparse matrix to a vector are tabulated in Table 1. The standard form of the operator A with $N = 256$ is depicted in Figure 5.

Input Size	Time			Error of Single Precision Multiplication		Error of FWT Multiplication		Compression Coefficient
N	T_s	T_w	T_d	L_2-norm	L_∞-norm	L_2-norm	L_∞-norm	C_{comp}
64	0.12	0.16	7.76	$1.26 \cdot 10^{-7}$	$3.65 \cdot 10^{-7}$	$8.89 \cdot 10^{-8}$	$1.72 \cdot 10^{-7}$	1.39
128	0.48	0.38	32.62	$2.17 \cdot 10^{-7}$	$8.64 \cdot 10^{-7}$	$1.12 \cdot 10^{-7}$	$9.94 \cdot 10^{-7}$	2.22
256	1.92	0.80	96.44	$2.81 \cdot 10^{-7}$	$1.12 \cdot 10^{-6}$	$1.25 \cdot 10^{-7}$	$5.30 \cdot 10^{-7}$	3.93
512	7.68	1.80	252.72	$4.21 \cdot 10^{-7}$	$1.75 \cdot 10^{-6}$	$1.23 \cdot 10^{-7}$	$5.16 \cdot 10^{-7}$	7.33
1024	30.72	3.72	605.74	$6.64 \cdot 10^{-7}$	$3.90 \cdot 10^{-6}$	$1.36 \cdot 10^{-7}$	$5.04 \cdot 10^{-7}$	14.09

TABLE 1. Numerical results for Example 4.2.

Column 1 of Table 1 contains the number N indicating the size of $N \times N$ matrix A_{ij}, columns 2, 3 contain CPU times T_s, T_w required by the standard order $O(N^2)$ and the fast $O(N)$ schemes to multiply a vector by the matrix, and column 4 contains the CPU T_d time used to produce the non-standard form of the operator. Columns 5, 6 contain the L_2 and L_∞ errors of the direct calculation, and columns 7, 8 contain the same information for the result obtained by computing in the wavelet system of coordinates. Finally, the last column contains the compression coefficients C_{comp}, defined by the ratio of N^2 to the number of non-zero elements in the non-standard form of of the matrix.

5. The operator d/dx in wavelet bases.

For a number of operators (e.g., differential operators, fractional derivatives, Hilbert and Riesz transforms) we may compute the non-standard form in the wavelet bases by solving a small system of linear algebraic equations [4]. As an example, we construct the non-standard form of the operator d/dx. The matrix elements α_{il}^j, β_{il}^j, and γ_{il}^j of A_j, B_j, and Γ_j, where $i, l, j \in \mathbf{Z}$ for the operator d/dx are easily computed as

$$(5.1) \qquad \alpha_{il}^j = 2^{-j} \int_{-\infty}^{\infty} \psi(2^{-j}x - i)\, \psi'(2^{-j}x - l)\, 2^{-j} dx = 2^{-j} \alpha_{i-l},$$

$$(5.2) \qquad \beta_{il}^j = 2^{-j} \int_{-\infty}^{\infty} \psi(2^{-j}x - i)\, \varphi'(2^{-j}x - l)\, 2^{-j} dx = 2^{-j} \beta_{i-l},$$

and

$$(5.3) \qquad \gamma_{il}^j = 2^{-j} \int_{-\infty}^{\infty} \varphi(2^{-j}x - i)\, \psi'(2^{-j}x - l)\, 2^{-j} dx = 2^{-j} \gamma_{i-l},$$

where

$$(5.4) \qquad \alpha_l = \int_{-\infty}^{+\infty} \psi(x - l)\, \frac{d}{dx} \psi(x)\, dx,$$

$$(5.5) \qquad \beta_l = \int_{-\infty}^{+\infty} \psi(x - l)\, \frac{d}{dx} \varphi(x)\, dx,$$

and

$$(5.6) \qquad \gamma_l = \int_{-\infty}^{+\infty} \varphi(x - l)\, \frac{d}{dx} \psi(x)\, dx.$$

Moreover, using (1.9) and (1.19) we have

$$(5.7) \qquad \alpha_i = 2 \sum_{k=0}^{L-1} \sum_{k'=0}^{L-1} g_k\, g_{k'}\, r_{2i+k-k'},$$

$$(5.8) \qquad \beta_i = 2 \sum_{k=0}^{L-1} \sum_{k'=0}^{L-1} g_k\, h_{k'}\, r_{2i+k-k'},$$

and

$$(5.9) \qquad \gamma_i = 2 \sum_{k=0}^{L-1} \sum_{k'=0}^{L-1} h_k\, g_{k'}\, r_{2i+k-k'},$$

where

$$(5.10) \qquad r_l = \int_{-\infty}^{+\infty} \varphi(x - l)\, \frac{d}{dx}\varphi(x)\, dx, \quad l \in \mathbf{Z}.$$

Therefore, the representation of d/dx is completely determined by the coefficients r_l in (5.10) or in other words, by the representation of d/dx on the subspace $\mathbf{V}_0$. Rewriting (5.10) in terms of $\hat{\varphi}(\xi)$ (see (1.11)), we obtain

$$(5.11) \qquad r_l = \int_{-\infty}^{+\infty} |\hat{\varphi}(\xi)|^2 (\mathrm{i}\xi) e^{-\mathrm{i}l\xi}\, d\xi.$$

Thus, the coefficients r_l depend only on the autocorrelation function of the scaling function φ, rather than the scaling function itself since the integral in (5.11) depends only on $|\hat{\varphi}(\xi)|^2$. The same holds, in fact, for all convolution operators [4].

Remark. The autocorrelation function of the scaling function (see (5.24)) has $2M - 1$ vanishing moments and its "zero moment" is equal to one (see (5.25) and (5.26)). This implies that if we consider the representation of the derivative operator on the subspace $\mathbf{V}_0$ as a finite-difference scheme, such a scheme has order $2M$. For integral convolution operators, this implies that the asymptotics is accurate to order $2M$ (see [4] and below).

The following observations [4] reduce the computation of the coefficients r_l to solving a system of linear algebraic equations.

1. If the integrals in (5.10) or (5.11) exist, then the coefficients r_l, $l \in \mathbf{Z}$ in (5.10) satisfy the following system of linear algebraic equations

$$(5.12) \qquad r_l = 2 \left[r_{2l} + \tfrac{1}{2} \sum_{k=1}^{L/2} a_{2k-1}(r_{2l-2k+1} + r_{2l+2k-1}) \right],$$

and

$$(5.13) \qquad \sum_l l\, r_l = -1,$$

where

$$(5.14) \qquad a_{2k-1} = 2 \sum_{i=0}^{L-2k} h_i\, h_{i+2k-1}, \quad k = 1, \ldots, L/2$$

are the autocorrelation coefficients of the filter H.

2. If $M \geq 2$, then the equations (5.12) and (5.13) have a unique solution with a finite number of non-zero r_l, namely, $r_l \neq 0$ for $-L + 2 \leq l \leq L - 2$ and

$$(5.15) \qquad r_l = -r_{-l},$$

Solving equations (5.12), (5.13), we present the results for Daubechies' wavelets with $M = 2, 3$. For further examples we refer to [**4**].

1. $M = 2$

$$a_1 = \frac{9}{8}, \quad a_3 = -\frac{1}{8},$$

and

$$r_1 = -\frac{2}{3}, \quad r_2 = \frac{1}{12},$$

We note that the coefficients $(-1/12, 2/3, 0, -2/3, 1/12)$ of this example can be found in many books on numerical analysis as a choice of coefficients for numerical differentiation.

2. $M = 3$

$$a_1 = \frac{75}{64}, \quad a_3 = -\frac{25}{128}, \quad a_5 = \frac{3}{128},$$

and

$$r_1 = -\frac{272}{365}, \quad r_2 = \frac{53}{365}, \quad r_3 = -\frac{16}{1095}, \quad r_4 = -\frac{1}{2920}.$$

The structure of non-standard and standard forms of derivative operators is illustrated in Figures 6 and 7.

FIGURE 6. Sparse structure of the non-standard form of derivative operators. The width of the bands depends only on the choice of the basis and is equal to $2L - 3$.

FIGURE 7. Sparse structure of the standard form of derivative operators.

For the coefficients $r_l^{(n)}$ of d^n/dx^n, $n > 1$, the system of linear algebraic equations is similar to that for the coefficients of d/dx. This system (and (5.12)) may be written in terms of

$$(5.16) \qquad \hat{r}(\xi) = \sum_l r_l^{(n)} e^{il\xi},$$

as

$$(5.17) \qquad \hat{r}(\xi) = 2^n \left(|m_0(\xi/2)|^2 \, \hat{r}(\xi/2) + |m_0(\xi/2 + \pi)|^2 \, \hat{r}(\xi/2 + \pi) \right),$$

where m_0 is the 2π-periodic function in (1.12). Considering the operator M_0 on 2π-periodic functions

$$(5.18) \qquad (M_0 f)(\xi) = |m_0(\xi/2)|^2 \, f(\xi/2) + |m_0(\xi/2 + \pi)|^2 \, f(\xi/2 + \pi),$$

we rewrite (5.17) as

$$(5.19) \qquad M_0 \hat{r} = 2^{-n} \hat{r},$$

so that $\hat{r}$ is an eigenvector of the operator M_0 corresponding to the eigenvalue 2^{-n}. Thus, finding the representation of the derivatives in the wavelet basis is equivalent to finding trigonometric polynomial solutions of (5.19) and vice versa [4].

An important property of the wavelet representation of the (periodized) derivative operators (and, in general, pseudodifferential operators with homogeneous symbols) is that these operators have an explicit diagonal preconditioner in wavelet bases.

We present here two tables illustrating such preconditioning applied to the standard form of the second derivative. In the following examples the standard form of the periodized second derivative $\mathbf{D}_2$ of size $N \times N$, where $N = 2^n$, is preconditioned in the wavelet basis by the diagonal matrix $\mathbf{P}$,

$$\mathbf{D}_2^p = \mathbf{P}\mathbf{D}_2^w\mathbf{P},$$

where $P_{il} = \delta_{il}2^j$, $1 \le j \le n$, and where j is chosen depending on i, l so that $N - N/2^{j-1} + 1 \le i, l \le N - N/2^j$, and $P_{NN} = 2^n$. The matrix $\mathbf{P}$ is illustrated in Figure 8.

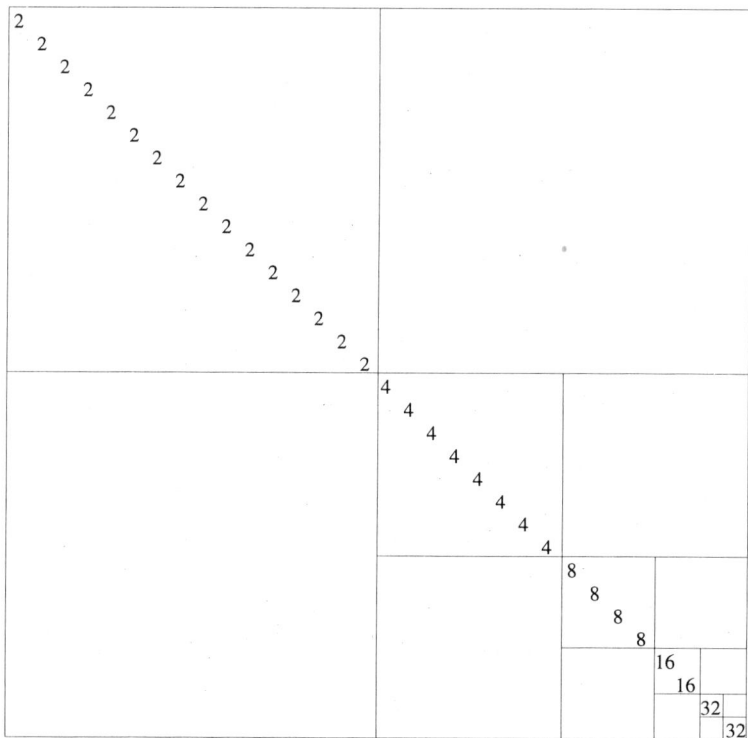

FIGURE 8. An example ($n = 5$) of the diagonal matrix $\mathbf{P}$ used to rescale the matrix of the periodized second derivative $\mathbf{D}_2^w$ in the wavelet system of coordinates.

Tables 2 and 3 below compare the original condition number κ of $\mathbf{D}_2^w$ and κ_p of $\mathbf{D}_2^p$.

Fractional derivatives. First, let us consider a convolution operator T and the infinite matrix $t_{i-l}^{(j-1)}$, $i, l \in \mathbf{Z}$, representing $P_{j-1}TP_{j-1}$ on the subspace $\mathbf{V}_{j-1}$.

N	κ	κ_p
64	0.14545E+04	0.10792E+02
128	0.58181E+04	0.11511E+02
256	0.23272E+05	0.12091E+02
512	0.93089E+05	0.12604E+02
1024	0.37236E+06	0.13045E+02

TABLE 2. Condition numbers of the matrix of the periodized second derivative (with and without preconditioning) in the basis of Daubechies' wavelets with three vanishing moments $M = 3$.

To compute the representation of $P_j T P_j$, we have (see e.g., formula (3.26) of [7])

$$(5.20) \qquad t_l^{(j)} = \sum_{k=0}^{L-1} \sum_{m=0}^{L-1} h_k \, h_m \, t_{2l+k-m}^{(j-1)}.$$

It easily reduces to

$$(5.21) \qquad t_l^{(j)} = t_{2l}^{(j-1)} + \frac{1}{2} \sum_{k=0}^{L/2} a_{2k-1} \left(t_{2l-2k+1}^{(j-1)} + t_{2l+2k-1}^{(j-1)} \right).$$

where the coefficients a_{2k-1} are given in (5.14).

We also have

$$(5.22) \qquad t_l^{(j)} = \int_{-\infty}^{+\infty} \int_{-\infty}^{+\infty} K(x-y) \, \varphi_{j,0}(y) \, \varphi_{j,l}(x) \, dxdy,$$

and, by changing the order of integration, we obtain

$$(5.23) \qquad t_l^{(j)} = 2^j \int_{-\infty}^{+\infty} K(2^j(l-y)) \, \Phi(y) \, dy,$$

where Φ is the autocorrelation function of the scaling function φ,

$$(5.24) \qquad \Phi(y) = \int_{-\infty}^{+\infty} \varphi(x) \, \varphi(x-y) \, dx.$$

It is easy to verify (see [4]) that

$$(5.25) \qquad \int_{-\infty}^{+\infty} \Phi(y) dy = 1,$$

N	κ	κ_p
64	0.10472E+04	0.43542E+01
128	0.41886E+04	0.43595E+01
256	0.16754E+05	0.43620E+01
512	0.67018E+05	0.43633E+01
1024	0.26807E+06	0.43640E+01

TABLE 3. Condition numbers of the matrix of the periodized second derivative (with and without preconditioning) in the basis of Daubechies' wavelets with six vanishing moments $M = 6$.

l	Coefficients r_l	l	Coefficients r_l
-7	-2.82831017E-06	4	-2.77955293E-02
-6	-1.68623867E-06	5	-2.61324170E-02
-5	4.45847796E-04	6	-1.91718816E-02
-4	-4.34633415E-03	7	-1.52272841E-02
-3	2.28821728E-02	8	-1.24667403E-02
-2	-8.49883759E-02	9	-1.04479500E-02
-1	0.27799963	10	-8.92061945E-03
0	0.84681966	11	-7.73225246E-03
1	-0.69847577	12	-6.78614593E-03
2	2.36400139E-02	13	-6.01838599E-03
3	-8.97463780E-02	14	-5.38521459E-03

TABLE 4. The coefficients $\{r_l\}_l$, $l = -7, \ldots, 14$ of the fractional derivative $\alpha = 0.5$ for Daubechies' wavelets with six vanishing moments.

and

$$(5.26) \qquad \mathcal{M}_\Phi^m = \int_{-\infty}^{+\infty} y^m \, \Phi(y) dy = 0, \quad \text{for} \quad 1 \le m \le 2M - 1.$$

The vanishing moments of the autocorrelation function Φ allow us to compute the elements of the matrix $t_l^{(j)}$ for large l and sufficiently fine scales $j \le 0$. Expanding the kernel K in its Taylor series, we obtain from (5.23)

$$(5.27) \; t_l^{(j)} = 2^j K(2^j l) + \frac{(-1)^{2M} 2^{(2M+1)j}}{(2M)!} \int_{-\infty}^{+\infty} K^{(2M)}(2^j(l - \tilde{y})) \, \Phi(y) \, dy,$$

where $\tilde{y} = \tilde{y}(y, l)$ and $K^{(2M)}$ denotes the $(2M)$th derivative of K. The decay of $K^{(2M)}(2^j(l - \tilde{y}))$ for large l is faster than that of the original kernel (see (4.1) and (4.2) with an appropriate choice of M) and (5.27) implies a one-point quadrature formula $t_l^{(j)} \approx 2^j K(2^j l)$ for large l and sufficiently fine scales $j \le 0$.

Computing representations of convolution operators simplifies further if the symbol of the operator is homogeneous of some degree. Let us illustrate this using the example of fractional derivatives. We define fractional derivatives as

$$(5.28) \qquad (\partial_x^\alpha f)(x) = \int_{-\infty}^{+\infty} \frac{(x - y)_+^{-\alpha-1}}{\Gamma(-\alpha)} f(y) dy,$$

where we consider $\alpha \ne 1, 2 \dots$. If $\alpha < 0$, then (5.28) defines fractional anti-derivatives.

The representation of ∂_x^α on $\mathbf{V}_0$ is determined by the coefficients

$$(5.29) \qquad r_l = \int_{-\infty}^{+\infty} \varphi(x - l) \, (\partial_x^\alpha \varphi)(x) \, dx, \quad l \in \mathbf{Z},$$

provided that this integral exists.

The non-standard form $\partial_x^\alpha = \{A_j, B_j, \Gamma_j\}_{j \in \mathbf{z}}$ is computed via $A_j = 2^{-\alpha j} A_0$, $B_j = 2^{-\alpha j} B_0$, and $\Gamma_j = 2^{-\alpha j} \Gamma_0$, where matrix elements α_{i-l}, β_{i-l}, and γ_{i-l} of A_0, B_0, and Γ_0 are obtained from the coefficients r_l,

$$(5.30) \qquad \alpha_i = 2^\alpha \sum_{k=0}^{L-1} \sum_{k'=0}^{L-1} g_k \, g_{k'} \, r_{2i+k-k'},$$

$$(5.31) \qquad \beta_i = 2^\alpha \sum_{k=0}^{L-1} \sum_{k'=0}^{L-1} g_k \, h_{k'} \, r_{2i+k-k'},$$

and

$$(5.32) \qquad \gamma_i = 2^\alpha \sum_{k=0}^{L-1} \sum_{k'=0}^{L-1} h_k \, g_{k'} \, r_{2i+k-k'}.$$

It easy to verify that the coefficients r_l satisfy the following system of linear algebraic equations

$$(5.33) \qquad r_l = 2^\alpha \left[r_{2l} + \frac{1}{2} \sum_{k=1}^{L/2} a_{2k-1}(r_{2l-2k+1} + r_{2l+2k-1}) \right],$$

where the coefficients a_{2k-1} are given in (5.14). Using (5.27), we obtain the asymptotics of r_l for large l,

$$(5.34) \qquad r_l = \frac{1}{\Gamma(-\alpha)} \frac{1}{l^{1+\alpha}} + O\left(\frac{1}{l^{1+\alpha+2M}}\right) \quad \text{for} \quad l > 0,$$

$$(5.35) \qquad r_l = 0 \qquad\qquad\qquad\qquad\qquad\quad \text{for} \quad l < 0.$$

EXAMPLE 5.1. We compute the coefficients r_l of the fractional derivative with $\alpha = 0.5$ for Daubechies' wavelets with six vanishing moments with accuracy 10^{-7}. The coefficients for r_l, $l > 14$ or $l < -7$ are obtained using the asymptotics

$$(5.36) \qquad r_l = -\frac{1}{2\sqrt{\pi}} \frac{1}{l^{1+\frac{1}{2}}} + O\left(\frac{1}{l^{13+\frac{1}{2}}}\right) \quad \text{for} \quad l > 0,$$

$$(5.37) \qquad r_l = 0 \qquad\qquad\qquad\qquad\qquad\quad \text{for} \quad l < 0.$$

6. Multiplication of matrices and fast iterative construction of the generalized inverse.

The standard and non-standard forms may be multiplied in fast manner if the matrices represent Calderón-Zygmund or pseudo-differential operators. Multiplication of matrices in the standard form is a straightforward algorithm [8], [1] and requires at most $O(N \log^2 N)$ operations. The algorithm for the multiplication of matrices in the non-standard form has been outlined in [3] and requires $O(N)$ operations. This is a significant improvement over $O(N^3)$ operations for dense matrices which arise in the ordinary discretization of the operators from these classes.

Fast multiplication algorithms give a second life to a great number of iterative algorithms. Indeed, powers of matrices may be computed as well as other functions of matrices. Let us consider an iterative construction of the generalized inverse. In order to construct the generalized inverse $A^\dagger$ of the matrix A, we use the following result [31]:

Let σ_1 be the largest singular value of the $m \times n$ matrix A. Consider the sequence of matrices X_k

$$(6.1) \qquad\qquad X_{k+1} = 2X_k - X_k A X_k$$

with

$$(6.2) \qquad\qquad X_0 = \alpha A^*,$$

where A^* is the adjoint matrix and α is chosen so that the largest eigenvalue of αA^*A is less than one. Then the sequence X_k converges to the generalized inverse $A^\dagger$.

Combining this iteration with fast multiplication algorithms, we obtain an algorithm for constructing the generalized inverse in at most $O(N \log^2 N \log R)$ operations, where R is the condition number of the matrix. (By the condition number we understand the ratio of the largest singular value to the smallest singular value above the threshold of accuracy).

The details of this algorithm (in the context of computing in wavelet bases) will be described in [9]. We note that throughout the iteration (6.1), it is necessary to maintain the "finger" band structure of the standard form of matrices X_k. Hence, the standard form of both the operator and its generalized inverse must admit such structure. We note that the pseudo-differential operators satisfy this condition.

Table 5 contains timings and accuracy comparisons for the construction of the generalized inverse via the singular value decomposition (SVD), which is an $O(N^3)$ procedure, and via the iteration (6.1)-(6.2) in the wavelet basis using the Fast Wavelet Transform (FWT). The computations were performed on a Sun Sparc workstation and we used a routine from LINPACK for computing the singular value decomposition. For tests we used the following full rank matrix

$$A_{ij} = \begin{cases} \frac{1}{i-j} & i \neq j \\ \\ 1 & i = j \end{cases},$$

where $i, j = 1, \ldots, N$. The accuracy threshold was set to 10^{-4}, i.e., entries of X_k below 10^{-4} were systematically removed after each iteration.

We note that the iteration in (6.1) also allows us to compute the projector on the null space (see [8] for this and several other examples).

The algorithm for the exponential is based on the identity

$$(6.3) \qquad \qquad \exp(A) = \left[\exp(2^{-L}A)\right]^{2^L}.$$

First, $\exp(2^{-L}A)$ is computed by means of the Taylor series (for instance). The number L is chosen so that the largest singular value of $2^{-L}A$ is less than one. At the second stage of the algorithm the matrix $\exp(2^{-L}A)$ is squared L times to obtain the result. Similarly, sine and cosine of a matrix can be computed using the elementary double-angle formulas. Unlike the algorithm for the generalized inverse, this algorithm is not self-correcting. Thus, it is necessary to maintain sufficient accuracy initially so as to obtain the desired accuracy after all the multiplications have been performed.

Size $N \times N$	SVD	FWT Generalized Inverse	L_2-Error
128×128	20.27 sec.	25.89 sec.	$3.1 \cdot 10^{-4}$
256×256	144.43 sec.	77.98 sec.	$3.42 \cdot 10^{-4}$
512×512	1,155 sec. (est.)	242.84 sec.	$6.0 \cdot 10^{-4}$
1024×1024	9,244 sec. (est.)	657.09 sec.	$7.7 \cdot 10^{-4}$
$\cdots$	$\cdots$	$\cdots$	$\cdots$
$2^{15} \times 2^{15}$	9.6 years (est.)	1 day (est.)	

TABLE 5. Comparison of the time needed to construct the generalized inverse of an $N \times N$ matrix via singular value decomposition (SVD) or the fast wavelet transform (FWT).

FIGURE 9. Standard form of the matrix $\mathbf{D}_b{}^{-1}$ computed via the iterative algorithm of this section with diagonal rescaling. Entries with absolute value greater than 10^{-8} are shown black.

Finally, as an example, let us consider the matrix

$$(6.4) \qquad \mathbf{D}_b = \begin{pmatrix} -2 & 1 & 0 & \cdots & 0 & 0 & 0 \\ 1 & -2 & 1 & \cdots & 0 & 0 & 0 \\ \cdots & \cdots & \cdots & \cdots & \cdots & \cdots & \cdots \\ 0 & 0 & 0 & \cdots & 1 & -2 & 1 \\ 0 & 0 & 0 & \cdots & 0 & 1 & -2 \end{pmatrix}.$$

which arises in the finite-difference formulation of the two-point boundary value problem. We note that the inverse of this matrix is sparse in the wavelet basis. As an illustration we display in Figure 9 the matrix $\mathbf{D}_b{}^{-1}$ obtained via the algorithm sketched above for computing the generalized inverse. Using the diagonal preconditioning (see Figure 8), this computation involves only well-conditioned matrices [5].

References

1. B. Alpert, G. Beylkin, R. R. Coifman, and V. Rokhlin, *Wavelet-like bases for the fast solution of second-kind integral equations*, SIAM Journal of Scientific and Statistical Computing **14** (1993), no. 1, 159–189, Technical report, Department of Computer Science, Yale University, New Haven, CT, 1990.
2. B. Alpert and V. Rokhlin, *A fast algorithm for the evaluation of Legendre expansions*, SIAM J. on Sci. Stat. Comput. **12** (1991), no. 1, 158–179, Yale University Technical Report, YALEU/DCS/RR-671 (1989).
3. G. Beylkin, *Wavelets, Multiresolution Analysis and Fast Numerical, Algorithms*, A draft of INRIA Lecture Notes (1991).
4. _____, *On the representation of operators in bases of compactly supported wavelets*, SIAM J. Numer. Anal. **29** (1992), no. 6, 1716–1740.
5. _____, *On wavelet-based algorithms for solving differential equations*, preprint (1992).
6. G. Beylkin and M. E. Brewster, *Fast numerical algorithms using wavelet bases on the interval*, in progress.
7. G. Beylkin, R. R. Coifman, and V. Rokhlin, *Fast wavelet transforms and numerical algorithms I*, Comm. Pure and Appl. Math. **44** (1991), 141–183, Yale University Technical Report YALEU/DCS/RR-696, August 1989.
8. _____, *Wavelets in Numerical Analysis*, Wavelets and Their Applications (M. B. Ruskai et al., ed.), Jones and Bartlett, 1992, pp. 181–210.
9. _____, *Fast wavelet transforms and numerical algorithms II*, in progress.
10. J. Carrier, L. Greengard, and V. Rokhlin, *A fast adaptive multipole algorithm for particle simulations*, SIAM Journal of Scientific and Statistical Computing **9** (1988), no. 4, Yale University Technical Report, YALEU/DCS/RR-496 (1986).
11. A. Cohen, I. Daubechies, B. Jawerth, and P. Vial, *Multiresolution analysis, wavelets and fast algorithms on an interval*, Comptes Rendus Acad. Sc. Paris, Série 1 **316** (1992), 417–421.
12. A. Cohen, I. Daubechies, and P. Vial, *Wavelets on the interval and fast wavelet transforms*, preprint (1993).
13. R. R. Coifman and Y. Meyer, *Nouvelles bases orthonormées de $l^2(\mathbf{R})$ ayant la structure du système de Walsh*, preprint (1989).
14. _____, *Remarques sur l'analyse de Fourier à fenêtre*, C.R. Acad. Sci. Paris, Série 1 **312** (1991), 259–261.
15. R. R. Coifman and V. Wickerhauser, *Best-adapted wave packet bases*, 1990.
16. I. Daubechies, *Orthonormal bases of compactly supported wavelets*, Comm. Pure and Appl. Math. **41** (1988), 909–996.
17. _____, *Ten lectures on wavelets*, CBMS-NSF Series in Applied Mathematics SIAM, 1992.

18. L. Greengard, *Potential flow in channels*, SIAM J. Sci. Stat. Comput. **11** (1990), no. 4, 603–620.
19. L. Greengard and V. Rokhlin, *A fast algorithm for particle simulations*, J. Comp. Phys. **73** (1987), no. 1, 325–348.
20. A. Haar, *Zur Theorie der orthogonalen Funktionensysteme*, Mathematische Annalen (1910), 331–371.
21. S. Jaffard, *Wavelet methods for fast resolution of elliptic problems*, SIAM Journal on Numerical Analysis **29** (1992), no. 4, 965–986.
22. A. Jouini and P.G. Lemarié-Rieusset, *Analyse multi-résolution biorthogonale sur l'intervalle et applications*, to appear in Ann. Inst. H. Poincaré.
23. S. Mallat, *Multiresolution approximation and wavelets*, Trans. Am. Math. Soc. **315** (1989), 69–88.
24. H. Malvar, *Lapped transforms for efficient transform/subband coding*, IEEE Trans. Acoust., Speech, Signal Processing **38** (1990), 677–680.
25. Y. Meyer, *Principe d'incertitude, bases hilbertiennes et algébres d'opérateurs*, Séminaire Bourbaki (Astérisque), Société Mathématique de France, Astérisque, 1985-86.
26. _____, *Ondelettes et fonctions splines*, Technical report, séminaire edp, Ecole Polytechnique, Paris, France, 1986.
27. _____, *Wavelets and operators*, Analysis at Urbana (E. Berkson N.T. Peck and J. Uhl, eds.), vol. 1, London Math. Society, 1989, Lecture Notes Series 137.
28. Yves Meyer, *Ondelettes et Opérateurs*, Hermann, Paris, 1990.
29. S.T. O'Donnel and V. Rokhlin, *A fast algorithm for the numerical evaluation of conformal mappings*, SIAM J. Sci. Stat. Comput. **10** (1989), no. 3, 475–487, Yale University Technical Report, YALEU/DCS/RR-554 (1987).
30. V. Rokhlin, *Rapid solution of integral equations of classical potential theory*, J. Comp. Phys. **60** (1985), no. 2.
31. G. Schulz, *Iterative Berechnung der reziproken Matrix*, Z. Angew. Math. Mech. **13** (1933), 57–59.
32. M. J. Smith and T. P. Barnwell, *Exact reconstruction techniques for tree-structured subband coders*, IEEE Transactions on ASSP **34** (1986), 434–441.
33. J. O. Stromberg, *A Modified Franklin System and Higher-Order Spline Systems on $\mathbf{R}^n$ as Unconditional Bases for Hardy Spaces*, Conference in harmonic analysis in honor of Antoni Zygmund, Wadworth math. series, 1983, pp. 475–493.

PROGRAM IN APPLIED MATHEMATICS, UNIVERSITY OF COLORADO AT BOULDER, BOULDER, CO 80309-0526

E-mail: beylkin@julia.colorado.edu

Proceedings of Symposia in Applied Mathematics
Volume **47**, 1993

Wavelets and Adapted Waveform Analysis.
A Toolkit for Signal Processing and Numerical
Analysis

RONALD R. COIFMAN AND M. VICTOR WICKERHAUSER

1. Introduction.

Wavelet analysis or more generally, Adapted Wave form Analysis (AWA) consists of a versatile collection of tools for the analysis and manipulation of signals such as sound and images, as well as more general digital data sets (including linear and non linear operators occurring in Simulations of Physical processes).

The term *adapted* relates to our ability to vary and match tools to given problems. These adaptive procedures can be achieved in a variety of ways depending on our goal. By analogy with music AWA provides us with the ability to represent a function or a signal in a mode similar to a musical score, where each note corresponds to a waveform having a duration pitch and amplitude. Our goal is to transcribe as efficiently as possible, and to orchestrate into different structures.

There are now in wavelet analysis two main modes for features and parameter extraction. The first and essentially the simplest follows the multiresolution scheme (Littlewood-Paley analysis) and consists in unraveling a signal, or an operator, according to scale, followed by an extraction of features, such as maxima or motion at each scale. These parameters can then be used for feature matching and reconstruction in various types of signal analysis, or as an analysis tool for unraveling the action of an operator on different scales and the interaction between scales. The second procedure matches a library of waveforms to a class of signals, such as speech, electrocardiograms, images, or to operators such as "acoustic potential" operators, and continues by finding, for each specific object, or class of signals, a most efficient representation in this library, resulting in high compression useful for storage and efficient computation.

1991 *Mathematics Subject Classification.* Primary 33C45, 42C10; Secondary 94A12.

Both approaches start by correlating the signal or image with a chosen collection of waveforms and continue through a selection of a small subset of parameters. Our goal is to introduce a variety of techniques permitting the mathematician, scientist or engineer to choose and apply appropriate analysis methods in this catalogue of tools. These methods correspond to the classical use of microlocalization in the study of Partial Differential Equations and Fourier Analysis, but rely on a computational approach to the subject.

By experimenting with the various software packages it is hoped that users can systematically fit parameter and structure extraction schemes to their goals.

In these lectures we will first concentrate on the second approach for parameter reduction in signal analysis, and relate these ideas to numerical analysis and operator study in the second part.

In the adapted waveform transform (AWT) method the user is provided with a collection of standard libraries of waveforms, (wavelet-packets, trigonometric waveforms) which can be chosen to fit specific classes of signals. These libraries come equipped with fast numerical algorithms, enabling realtime implementation of a variety of signal processing tasks such as data compression, extraction of parameters for recognition and diagnostics, as well as fast transformation and manipulation of digital data.

The process of analysis is usually done by comparing acquired segments of data with stored waveforms. The most efficient orthonormal basis for compression of the signal is selected and used to extract and manipulate relevant features.

Current applications permit good quality compressions of voice at 2-3 kbs, as well as high quality image compressions with small degradation of textures. Other applications include tools for denoising and enhancement of signals, fast numerical algorithms for computations for statistical recognition and diagnostic analysis, promising real time devices.

A calculus in compressed variables has been developed enabling the implementation of adapted transform methods for fast numerical algorithms useful for data manipulation and for large scale computation.

2. Windowed FFT and adapted window selection.

We start with a description of an algorithm to compute the Fourier expansion of a function on a union of two adjacent intervals of the same size in terms of the Fourier expansion on each interval.

Let f be defined on $[0, 2]$

$$f = f^0 + f^1 \qquad \text{where } f^0 = \begin{cases} f & x \in [0, 1] \\ 0 & x \notin [0, 1] \end{cases} ;$$

we want to compute

$$\hat{f}_m = \frac{1}{\sqrt{2}} \int_0^2 f(t) e^{-2\pi i m \frac{t}{2}} dt$$

in terms of $\hat{f}_n^0 = \int_0^1 f(t)e^{-2\pi int}dt$ and

$$\hat{f}_n^1 = \int_1^2 f(t)e^{-2\pi int}dt.$$

Clearly, when $m = 2n$ we have

$$\boxed{\hat{f}_{2n} = \frac{1}{\sqrt{2}}\{\hat{f}_n^0 + \hat{f}_n^1\}}.$$

For $m = 2n + 1$ we define

$$d_n = \frac{1}{\sqrt{2}}\{\hat{f}_n^0 - \hat{f}_{n+1}^1\}$$

and find

$$\boxed{\hat{f}_{2n+1} = \frac{1}{\pi i}\sum \frac{d_k}{(n-k+\frac{1}{2})}}.$$

In fact,

$$\hat{f}_{2n+1} = \frac{1}{\sqrt{2}}\int_0^1 [f(t) - f(t+1)]e^{-i\pi t}e^{-2\pi int}dt.$$

Since d_n are the Fourier coefficients on $[0,1]$ of $f(t) - f(t+1)$, and $\frac{1}{\pi i(n+\frac{1}{2})}$ are the coefficients of $e^{-i\pi t}$, we obtain the coefficients of $\hat{f}_{2n+1}$ by convolving these sequences.

Another way to compute $\hat{f}_{2n+1}$ is to compute the inverse transform on $(0,1)$ of d_n, multiply by $e^{-it/2}$ and recompute the transform on $(0,1)$.

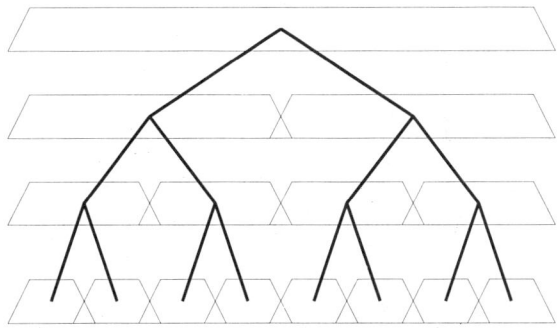

FIGURE 1. Schematic description.

We see that in order to compute the transform on the large interval, we can start with adjacent pairs of small intervals, combine coefficients to obtain the expansion on their union, and continue until we reach the top level. (Figure 1 gives a schematic description of this procedure.) As a result we have obtained all dyadic windowed Fourier transform as intermediate computations. Clearly every disjoint collection of intervals equipped with an orthogonal basis on each interval provides us with an orthogonal basis for the union. A natural question that arises in connection with the windowed Fourier transform is how to place the windows (see Figures 2 and 3 where the effect of the window selection on the number of large coefficients is visible).

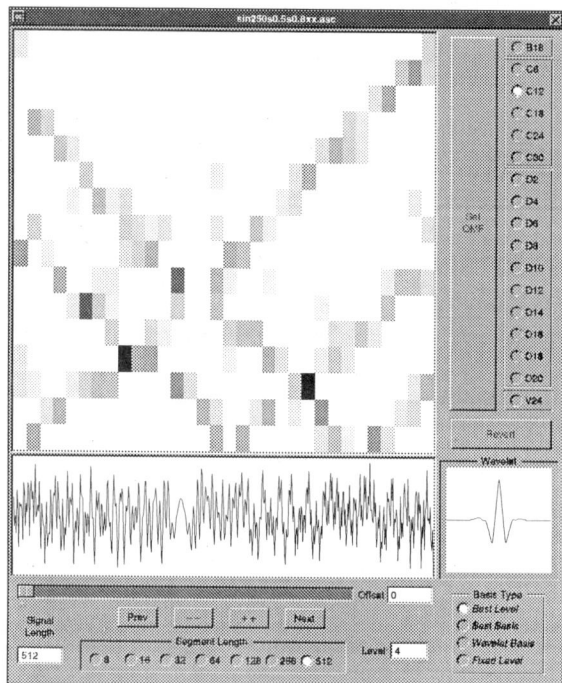

FIGURE 2. Optimal window selection.

For the moment let us consider the question of optimizing the windows to obtain an efficient representation of a function. We can proceed as follows: We start with the adjacent small intervals and consider the expansion coefficient in each separately. We then compute the expansion coefficients on their union. We can now choose that expansion for which the number of coefficients needed to capture 99% of the energy is smallest (or that expansion whose "cost" is smallest; e.g. information cost, coding cost, error cost). We compare the cost of the chosen expansions on two adjacent unions of pairs to the expansion on their union and again pick the best. We continue until we reach an optimal distribution of windows (see Figure 4 where the windows were adapted to a voice recording).

The procedure described above, although natural, is not very useful if we take the windowed Fourier transform with discontinuous windows, since the discontinuity introduces "large" expansion coefficients. A cosine basis on each interval is somewhat better. On the other hand, it is well known that we cannot find a smooth window function $\omega(x)$ supported on $(-\frac{1}{2}, \frac{3}{2})$ such that $\omega(x-k)e^{2i\pi nx}$

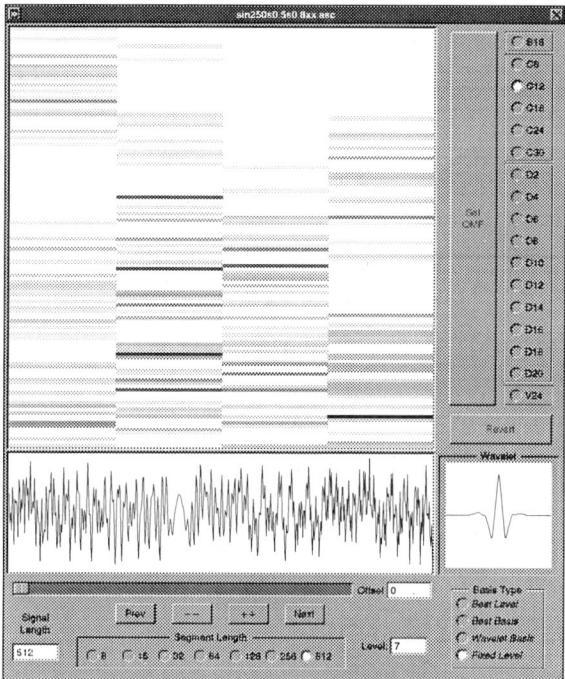

FIGURE 3. Large window selection.

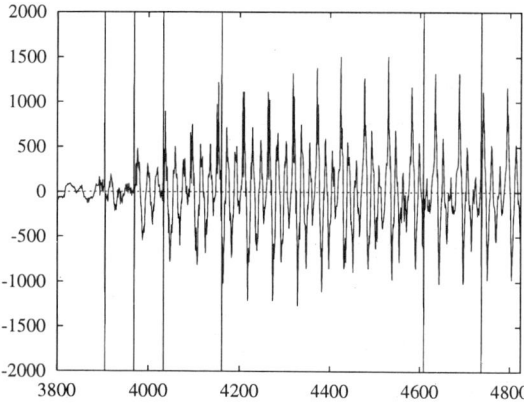

FIGURE 4. Adapted window selection.

are orthogonal. (This would imply $\int \omega(x)\omega(x-1)e^{-2\pi imx}dx = 0$ for all m i.e. $\omega(x)\omega(x-1) = 0$).

Recently Daubechies, Jaffard and Journé (1991) and Malvar (1990) observed that orthogonality can be maintained by taking equal windows and alternating sines or cosines. Coifman and Meyer (1991) observed that the windows can be chosen of different sizes enabling adaptive constructions as above.

3. Modulated waveform libraries.

We start by recalling the concept of a "Library of orthonormal bases". For the sake of exposition we restrict our attention to two classes of numerically useful waveforms, introduced recently (see Coifman, Meyer and Wickerhauser (1992) and Coifman and Meyer (1991)).

We start with trigonometric waveform libraries. These are localized sine transforms LST associated to a covering by intervals of $\mathbf{R}$ (more generally, of a manifold).

We consider a cover $\mathbf{R} = \bigcup_{-\infty}^{\infty} I_i$, where $I = [\alpha_i \alpha_{i+1})$ with $\alpha_i < \alpha_{i+1}$. Write $\ell_i = \alpha_{i+1} - \alpha_i = |I_i|$ and let $p_i(x)$ be a window function supported in $[\alpha_i - \ell_{i-1}/2, \alpha_{i+1} + \ell_{i+1}/2]$ such that

$$\sum_{-\infty}^{\infty} p_i^2(x) = 1$$

and

$$p_i^2(x) = 1 - p_i^2(2\alpha_{i+1} - x) \quad \text{for} \quad x \quad \text{near} \quad \alpha_{i+1} .$$

Then the functions

$$S_{i,k}(x) = \frac{2}{\sqrt{2\ell_i}} p_i(x) \sin[(2k+1)\frac{\pi}{2\ell_i}(x - \alpha_i)] ,$$

where $i \in \mathbb{Z}$, $k \in \mathbb{N}$, form an orthonormal basis of $L^2(\mathbf{R})$ subordinate to the partition p_i. The collection of such bases forms a library of orthonormal bases, Coifman and Meyer (1991). Figure 5 shows the graph of a few $S_{i,k}$ corresponding to $\alpha_i = -.5$, $\alpha_{i+1} = .5$. (In fact, the graph shows localized cosine basis functions, which can also be viewed as the mirror images of $S_{i,k}(x)$ with respect to $x = 0$.)

It is easy to check that if H_{I_i} denotes the space of functions spanned by $S_{i,k}$, $k = 0, 1, 2, ...$ then $H_{I_i} + H_{I_{i+1}}$ is spanned by the functions

$$P(x)\frac{1}{\sqrt{2(\ell_i + \ell_{i+1})}} \sin[(2k+1)\frac{\pi}{2(\ell_i + \ell_{i+1})}(x - \alpha_i)]$$

where

$$P^2 = p_i^2(x) + p_{i+1}^2(x)$$

is a "window" function covering the interval $I_i \cup I_{i+1}$.

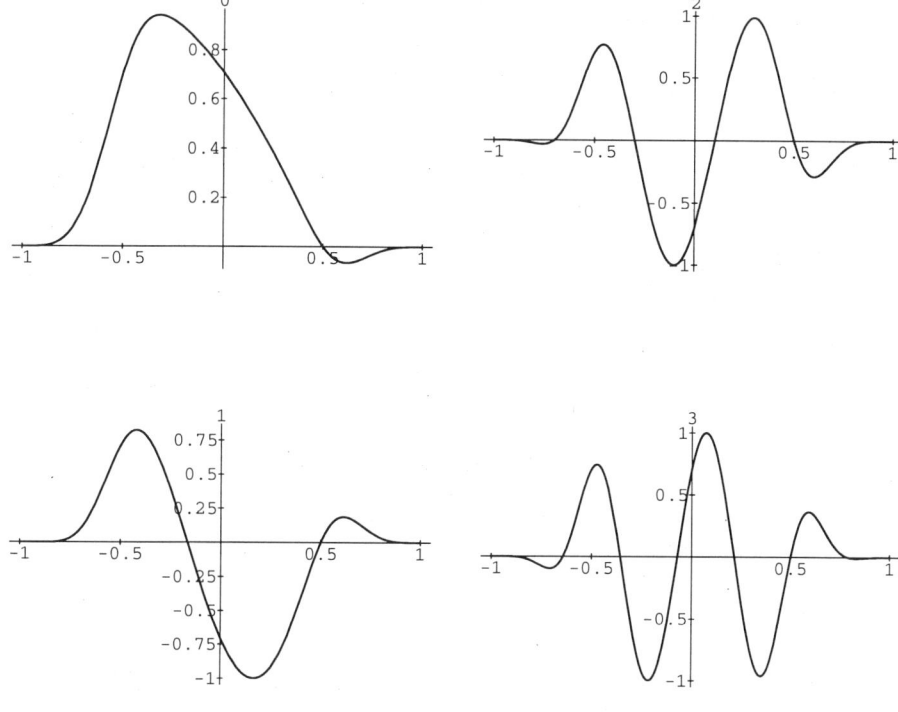

FIGURE 5. Examples of functions in an LCT basis.

3.1. Relation to wavelets – Wavelet packets. We consider the frequency line $\mathbf{R}$ split as $\mathbf{R}^+ \cup \mathbf{R}^-$, with $\mathbf{R}^+ = (0, \infty)$ and $\mathbf{R}^- = (-\infty, 0)$. On $L^2(0, \infty)$ we introduce a window function $p(\xi)$ such that $\sum_{k=-\infty}^{\infty} p^2(2^{-j}\xi) = 1$ and $p(\xi)$ is supported in $(3/4, 3)$. Clearly we can view $p(2^{-j}\xi)$ as a window function above the interval $(2^j, 2^{j+1})$ and observe that the functions

$$ s_{j,k}(\xi) = 2^{-j/2} \sin\left[\left(k + \frac{1}{2}\right) \pi \left(\frac{\xi - 2^j}{2^j}\right)\right] p(2^{-j}\xi), \quad j, k \in \mathbb{Z}, $$

form an orthonormal basis of $L^2(\mathbf{R}^+)$. Similarly $c_{j,k} = \cos\left[\left(k + \frac{1}{2}\right) \pi \left(\frac{\xi - 2^j}{2^j}\right)\right]$ $p(2^{-j}\xi)$ gives another basis, whose elements are not orthogonal. If we define $S_{k,j}$ as the odd extension to $\mathbf{R}$ of $s_{j,k}$ and $C_{j,k}$ as the even extension of $c_{k,j}$ we find $S_{j,k} \perp C_{j',k'}$ permitting us to write $C_{j,k} \pm i S_{j,k} = e^{\pm i k \pi \xi / 2^j} \hat{\psi}(\xi/2^j)$ where $\hat{\psi}(\xi) = e^{i\pi/2\xi} p(\xi)$ is the Fourier transform of the wavelet Ψ (see Meyer). We therefore see that wavelet analysis corresponds to windowing frequency space in "octave" windows $(2^j, 2^{j+1})$.

A natural extension is therefore provided by allowing all dyadic windows in frequency space and adapting the window choice by regrouping at will. This sort of analysis is equivalent to wavelet packet analysis.

The wavelet packet analysis algorithms permit us to perform an adapted

Fourier windowing directly in the time domain by successive filtering of a function into different regions in frequency. The dual version of the window selection provides an adapted subband coding algorithm.

The wavelet packet library is constructed by iterating the wavelet algorithm. This library contains the wavelet basis, Walsh functions, and smooth versions of Walsh functions called wavelet packets (Coifman, Meyer and Wickerhauser (1992)).

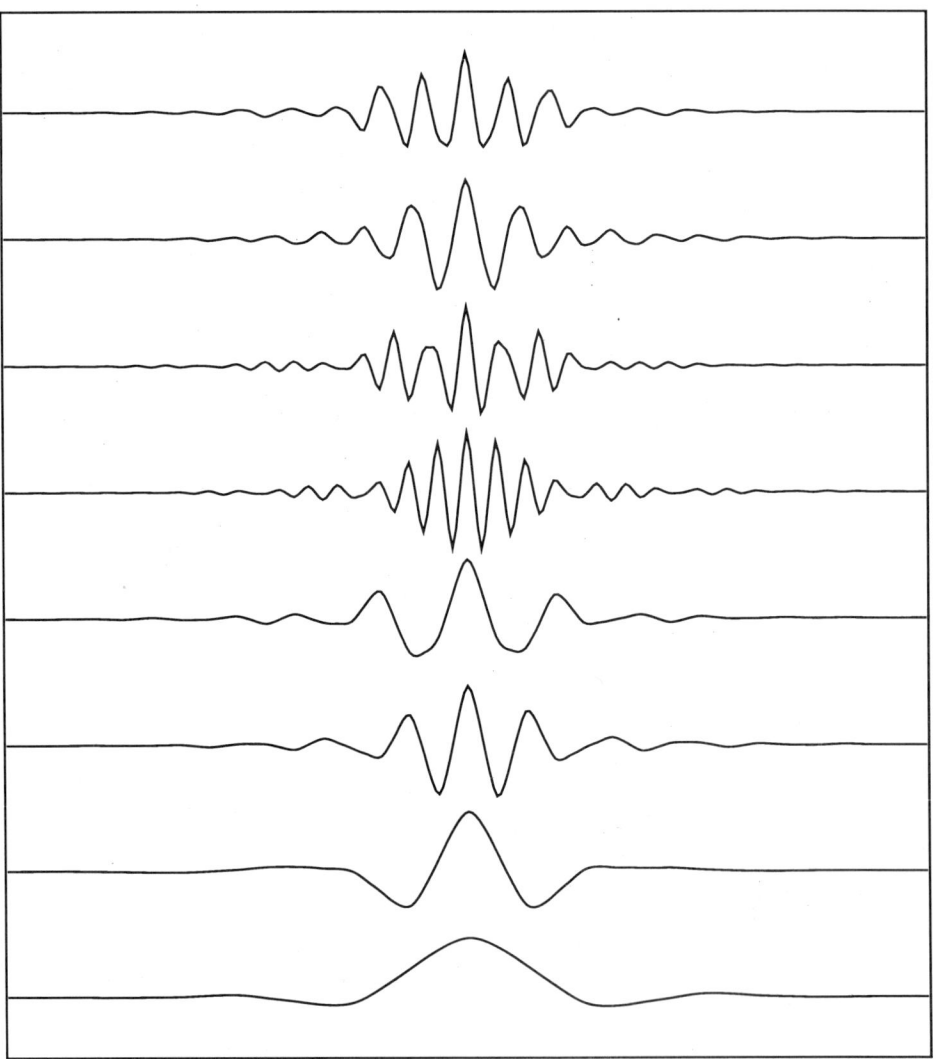

FIGURE 6. Wavelet-packets.

These waveforms are mutually orthogonal, moreover, each of them is orthogonal to all of its integer translates and dyadic rescaled versions. The full collection of these wavelet packets (including translates and rescaled versions) provides us with a library of "templates" or "notes" which are matched "efficiently" to signals for analysis and synthesis, Coifman, Meyer and Wickerhauser (1992), Coifman and Wickerhauser (1992), Coifman, Meyer, Quake and Wickerhauser (1992). Wavelet packet expansions correspond algorithmically to subband coding schemes and are numerically as fast as the FFT.

One can measure how good the fit is by introducing a notion of "distance" between a basis and a function in terms of the Shannon entropy of the expansion. More generally, let H be a Hilbert space; let $v \in H$, $\|v\| = 1$ and assume that H decomposes into an orthogonal direct sum,

$$H = \oplus \sum H_i .$$

We define

$$\varepsilon^2(v, \{H_i\}) = - \sum \|v_i\|^2 \ell n \|v_i\|^2$$

as a measure of distance between v and the orthogonal decomposition. In particular, if v already lies in one of the H_i, then $\varepsilon^2(v, \{H_i\}) = 0$, meaning that the decomposition is optimal into 1 nonzero component, all other components zero. If v has nonzero components in several H_i, then $\varepsilon^2(v, \{H_i\}) > 0$. ε^2 is characterized by the Shannon equation which is a version of Pythagoras' theorem. Indeed, if

$$\begin{aligned} H &= \oplus(\sum H^i) \oplus (\sum H_j) \\ &= H_+ \oplus H_- , \end{aligned}$$

where H^i and H_j give orthogonal decompositions $H_+ = \sum H^i, H_- = \sum H_j$, then

$$\begin{aligned} \varepsilon^2(v; \{H^i, H_j\}) &= \varepsilon^2(v, \{H+, H_-\} + \|v_+\|^2 \varepsilon^2 \left(\frac{v_+}{\|v_+\|}, \{H^i\} \right) \\ &+ \|v_-\|^2 \varepsilon^2 \left(\frac{v_-}{\|v_-\|}, \{H_j\} \right) . \end{aligned}$$

This is Shannon's equation for entropy (if we interpret, as in quantum mechanics, $\|P_{H_+} v\|^2$ as the "probability" of v to be in the subspace H_+).

This equation enables us to search for a smallest entropy expansion of a signal. For example in the LST Library case, we compare the entropy of the expansion in two adjacent windows to the entropy of the expansion on their union, and pick the least expensive of the two possibilities, continuing the comparison with the selection made for the next pair, etc. (see Figure 7).

4. Time-frequency analysis.

To each wavelet packet or local trigonometric function we can associate a time t and a frequency f. These will be uncertain by amounts Δt and Δf, respectively. The result may be interpreted as a rectangular patch of dimensions Δt by Δf,

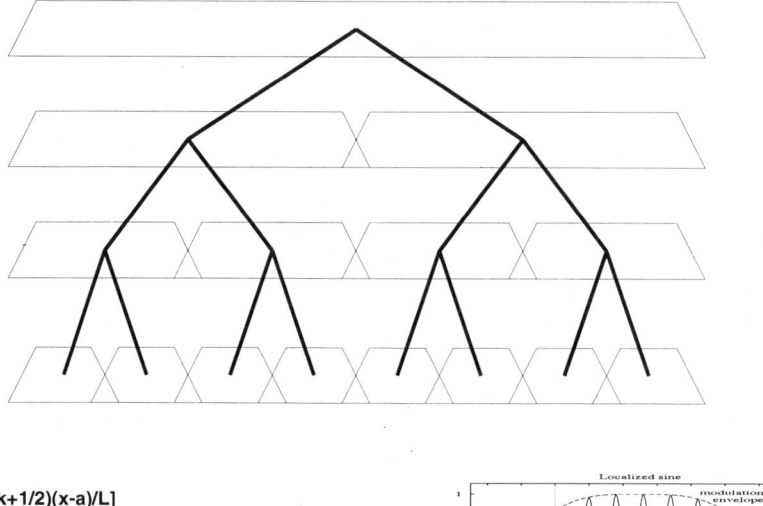

b(x) sin[(k+1/2)(x-a)/L]

a - endpoint ,
L -length

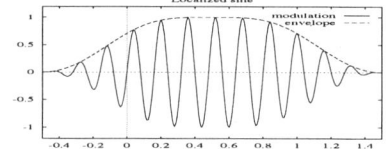

FIGURE 7. Tree search in the LST library.

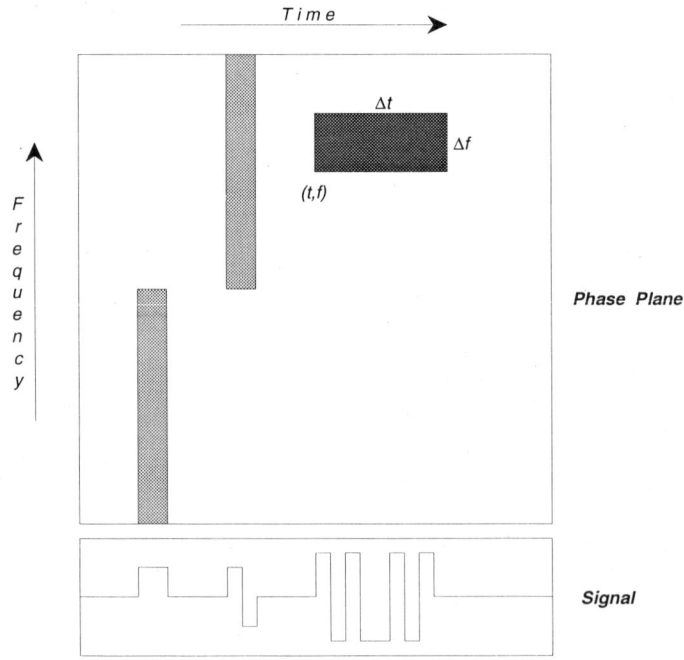

FIGURE 8. Cells in the phase plane.

located around (t, f). We shall call the patch a *phase cell* or Heisenberg box in honor of the uncertainty principle, which limits how small the area of the patch may be. The cells may be colored in proportion to the amplitude of the corresponding wavelet packet component.

An orthonormal basis corresponds to a disjoint cover of the phase plane by phase cells (Heisenberg boxes). Certain bases have characterizations in terms of

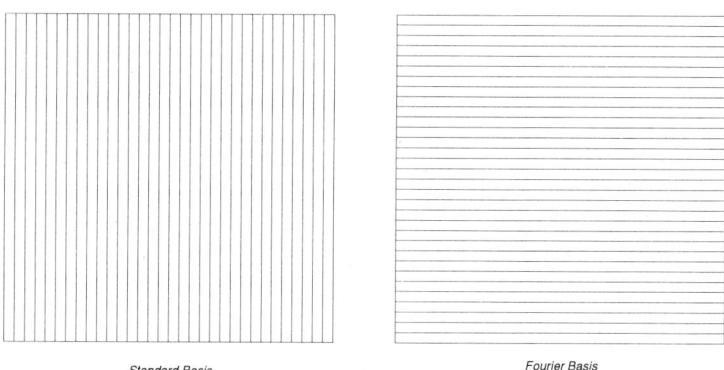

Standard Basis   Fourier Basis

FIGURE 9. Phase plane decomposition by the standard and Fourier bases.

the shapes of the boxes present in the cover. For example, the standard basis consists of the cover by the tallest, thinnest patches allowed by the sampling interval. The Fourier transform may be regarded as the transpose of the standard basis, in the sense that the cells are transposed by interchanging time and frequency. The standard basis has optimal time localization and no frequency localization, while the Fourier basis has optimal frequency localization, but no time localization (see Figure 9).

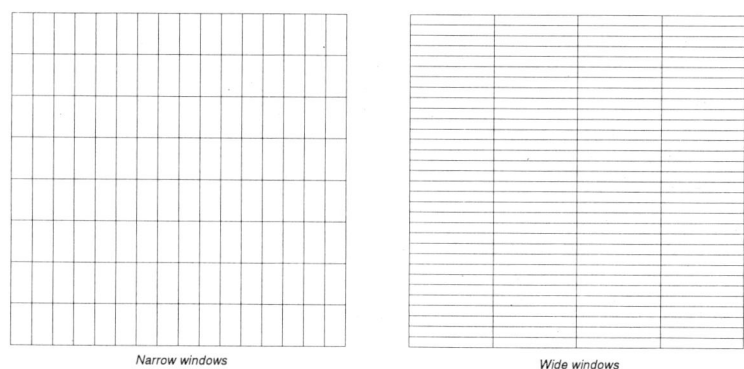

Narrow windows   Wide windows

FIGURE 10. Phase plane decomposition by windowed cosine transforms.

Windowed Fourier or cosine transforms with a fixed window size correspond to covers with congruent cells whose width Δt is the window width. The ratio of frequency uncertainty to time uncertainty is the aspect ratio of the cells; two different examples are shown in Figure 10.

The wavelet basis is an octave-band decomposition of the phase plane, depicted by the following cover:

Wavelet basis

FIGURE 11. Phase plane decomposition by wavelet transform.

The best-basis of wavelet packets fits a cover to the signal so as to minimize the amount of dark phase cell boxes. The compressibility of a sampled signal is easily seen to be the ratio of the total area of the phase plane ($N \times N$ for a signal sampled at N points) divided by the total area of the dark cells (each of area N). This method allows rectangles of all aspect ratios. The best-level or adapted subband basis fits a cover of equal aspect ratio rectangles to the signal, so as to minimize the amount of dark. We may automatically analyze signals by expanding them in the best basis, then drawing the corresponding phase plane representation. As is clear, the negligible components will not be drawn, as it is not relevant which particular basis is chosen for a subspace containing negligible energy.

Below are certain canonical signals and their automatic analyses by a wavelet packet program written for a desktop computer. The user selects a quadrature mirror filter from a list of 17 at the right, and the "mother wavelet" determined by that filter is displayed in the small square window at the lower right. The signal is plotted in the rectangular window at bottom, and the phase plane representation is drawn in the large main square window.

We first analyze a relatively smooth transient, spread over 7 samples in a 512 sample signal; see Figure 12.

Notice that the wavelet analysis at the right correctly localizes the peak in the high-frequency components, but is forced to include poorly localized low-frequency elements as well. The best-basis analysis finds the optimal represen-

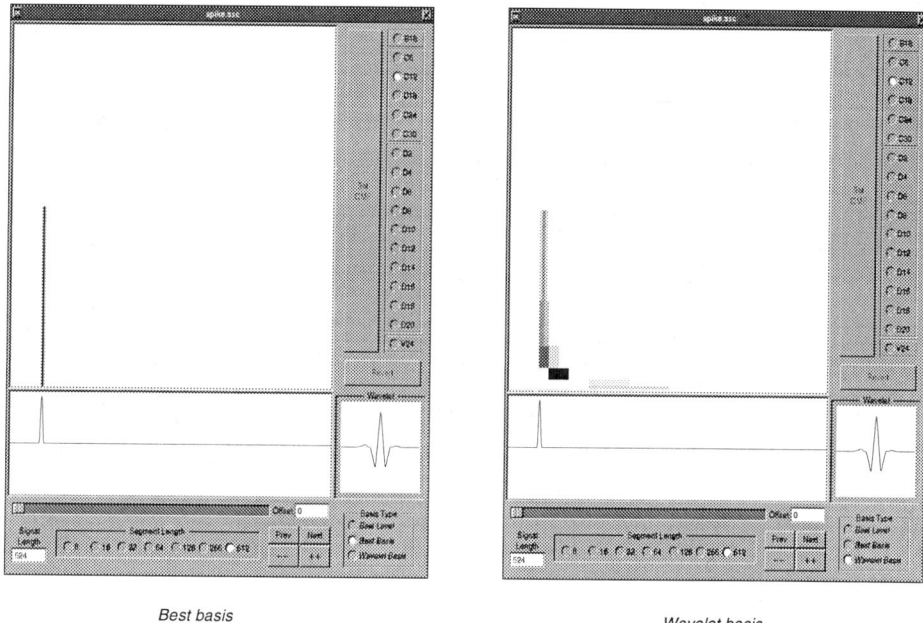

Best basis   Wavelet basis

FIGURE 12. Representing a fast transient.

tation within the library, which in this case is almost a single wavelet packet.

The second signal is taken from a recording (at 8012 samples per second) of a person whistling, shown in Figure 13.

Here the wavelet basis is only able to localize the frequency within an octave, even though the best-basis analysis shows that it falls in a much narrower band. The vertical stripes among the wavelet Heisenberg boxes may be used to further localize the frequencies, but the best-basis decomposition performs this analysis automatically.

Let us now combine the transient and periodic parts in different ways. For example, we may take a critically damped oscillator which receives an impulse, and decompose the resulting solution in the wavelet and best-level bases, as below. The wavelet decomposition locates the discontinuity at the impulse, while the best-level analysis finds the resonant frequency of the oscillator more precisely. The exponential decay of the amplitude is visible in both analyses.

A chirp is an oscillatory signal with increasing modulation. For example, below are the functions $\sin(250\pi t^2)$ and $\sin(190\pi t^3)$ on the interval $0 < x < 1$, sampled 512 times. The modulation increases linearly and quadratically, respectively. The Heisenberg boxes form a line and a parabolic arc, respectively. In the best-level analyses, all the Heisenberg boxes have the same aspect ratio, which is appropriate for a line. In the best-basis analysis, the Heisenberg boxes near the zero-slope portion have smaller aspect ratio than those near the large-slope portion.

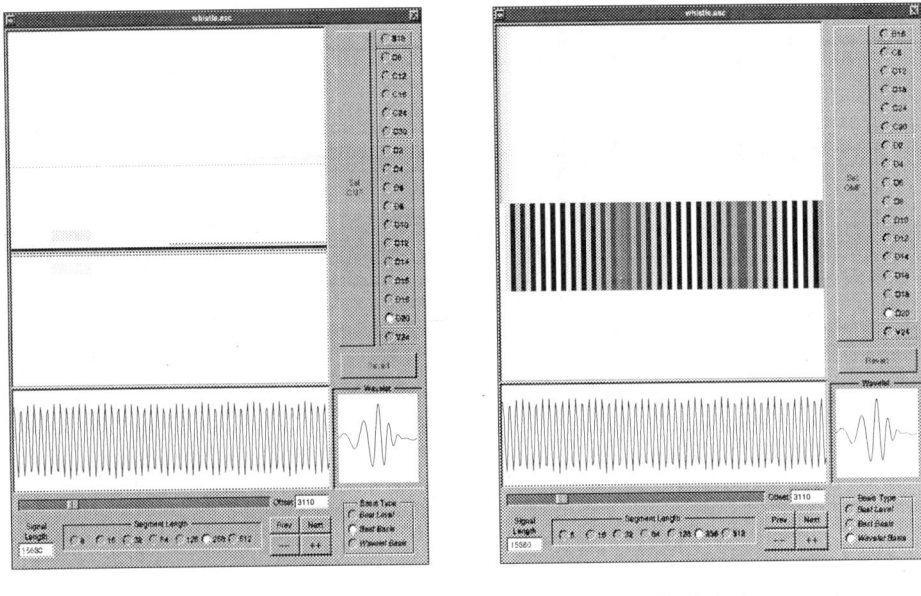

Best basis   Wavelet basis

FIGURE 13. Representing a whistle.

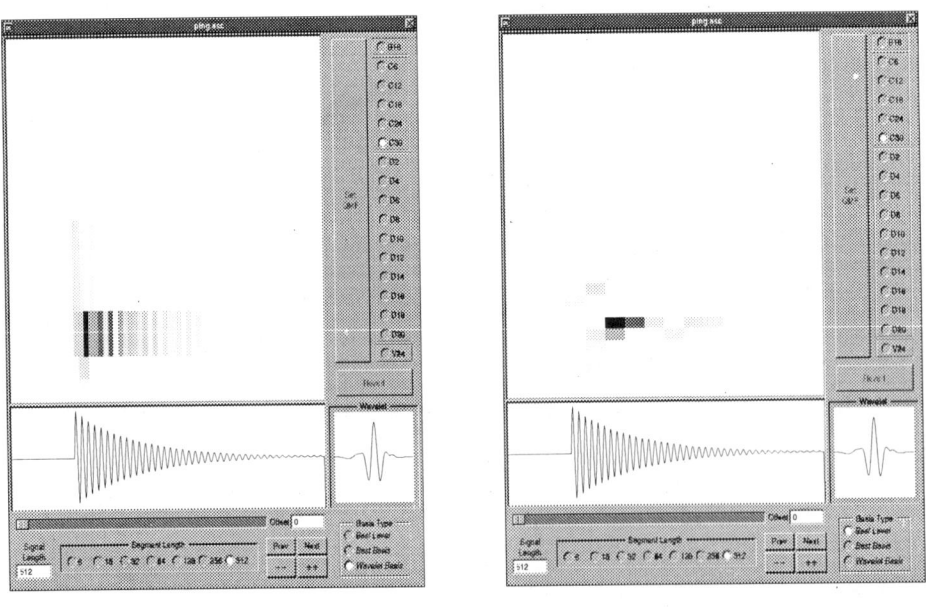

Wavelet basis   Best-level basis

FIGURE 14. Critically damped harmonic oscillator.

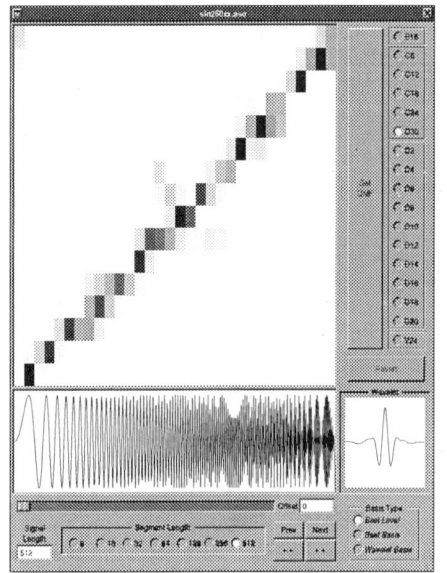

Linear chirp, best level

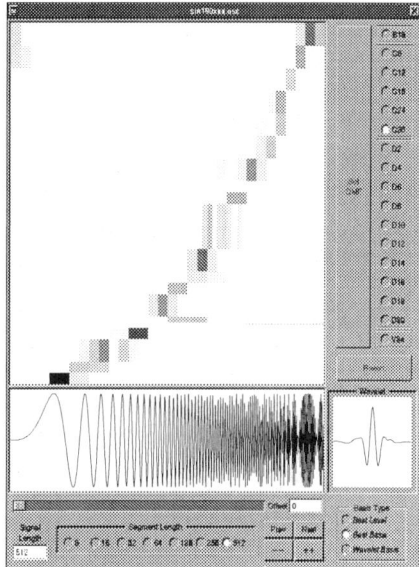

Quadratic chirp, best basis

FIGURE 15. Linear and quadratic chirps.

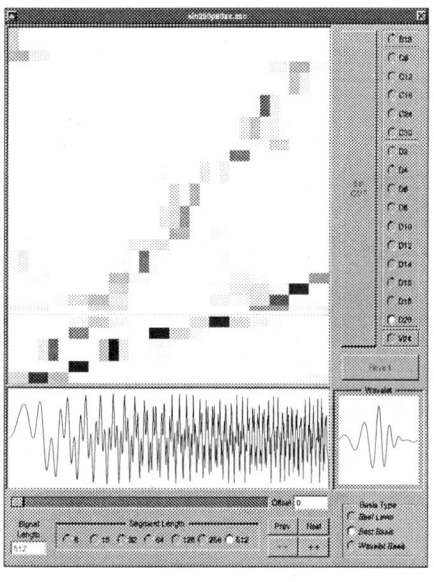

Different slopes, best basis

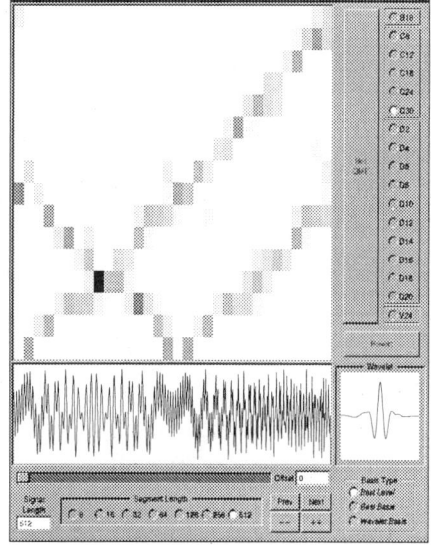

Different phases, best level

FIGURE 16. Superposed chirps.

Such a time-frequency analysis can separate superposed chirps. Below are pairs of linear chirps, differing either by modulation law or phase. Both are functions on the interval $0 < t < 1$, sampled 512 times. On the left is the function $\sin(250\pi t^2) + \sin(80\pi t^2)$ analyzed in the best wavelet packet basis. Note that the milder slope chirp is represented by Heisenberg boxes of lower aspect ratio. On the right is $\sin(250\pi t^2) + \sin(250\pi(t - \frac{1}{2})^2)$, analyzed by best-level wavelet packets. The downward-sloping line comes from the aliasing of negative frequencies.

5. Multidimensional local trigonometric transforms.

As seen previously we can partition the x axis into windows with smooth overlap while maintaining orthogonality of appropriate cos or sin wave forms.

A similar procedure is possible in higher dimensions (say 2) either by simply taking products of the basis in x with a basis in y or, more generally, by segmenting into rectangles or squares chosen appropriately. The only consideration is to assure compatibility of the various window functions.

For simplicity we can start by considering a basis of the form $S_{i,K}(x) = \sqrt{2}p(x - i)\sin[(2k + 1)\frac{\pi}{2k}(x - i)]$ where $\sum p^2(x - i) = 1$ and p is a bell above $[0, 1]$ such that $p^2(x) = 1 - p^2(-x)$ near 0, $1 - p^2(2 - x) = p^2(x)$ near 1 and a similar basis $S_{i',k'}(y)$. Clearly the functions

$$S_{i,K}(x)S_{i',K'}(y) = S_{(i,i'),(K,K')}(x, y)$$

form a local trigonometric basis in $\mathbf{R}^2$, based on squares of "size" 1. Assume now that we expand a function on a square of size 2×2 in terms of this basis based on four subsquares of size 1 and compare the "cost" of this expansion to the cost on the parent square, thereby choosing the most efficient expansion. Such a procedure can continue for several generations leading to a choice of basis based on squared of different sizes, as described in the following fingerprint analysis. (Here the bell function chosen for the square $(0, 2) \times (0, 2)$ would be $P^{(2)}(x, y) = \sqrt{(p^2(x) + p^2(x - 1))(p^2(y) + p^2(y - 1))}$ where the windows are picked by efficiency in coding, or approximation. See the following figures where the selected windows are superposed, indicating the best choice of LST; Figure 17 shows a fingerprint, Figure 18 the entries of a large matrix with oscillatory entries (rendered as an image: large positive entry = black, large negative entry = white).

This adapted box expansion for an operator involved in the computation of acoustic scattering in two dimensions provides an efficient way to describe explicitly the interactions between oscillations and geometry; the compressed version is in effect a fast numerical algorithm for applying this image viewed as a matrix to a vector.

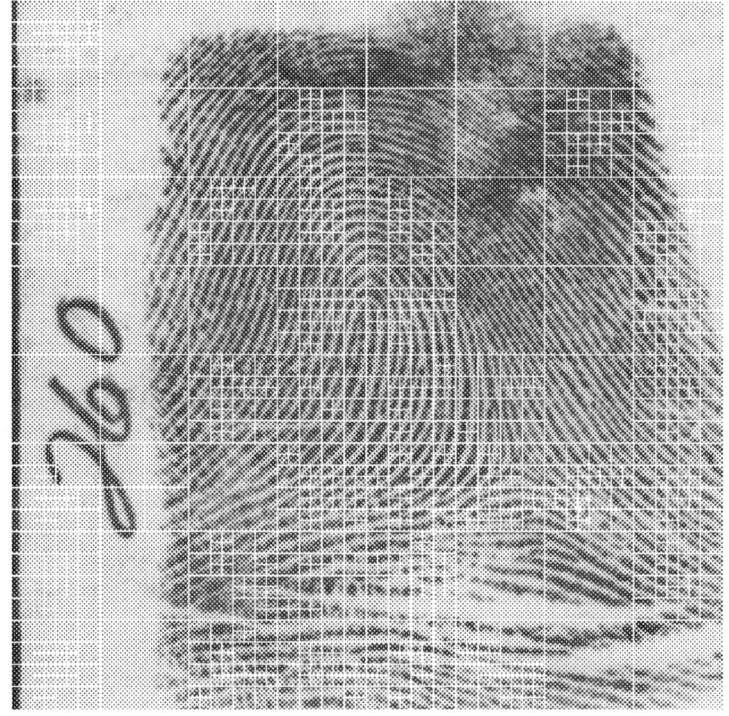

FIGURE 17. LST window selection.

6. The Haar system algorithms.

The Haar wavelet is defined by

$$h(x) = \begin{cases} 1 & 0 < x \leq \frac{1}{2} \\ -1 & \frac{1}{2} < x \leq 1 \\ 0 & x \leq 0 \text{ or } 1 < x \end{cases}$$

The Haar basis functions are rescaled versions of $h(x)$ (by 2^j) shifted by $2^{-j}k$,

$$h_k^j(x) = 2^{j/2}h(2^j x - k) \qquad \begin{matrix} j = 0, \pm 1, \pm 2 \dots \\ k = 0, \pm, \pm 2 \dots \end{matrix}$$

These functions are orthogonal i.e.

$$\langle h_k^j, h_{k'}^{j'} \rangle \equiv \int h_k^j(x) h_{k'}^{j'}(x) dx = \begin{cases} 1 & \text{if } j = j' \quad k = k' \\ 0 & \text{otherwise} \end{cases}$$

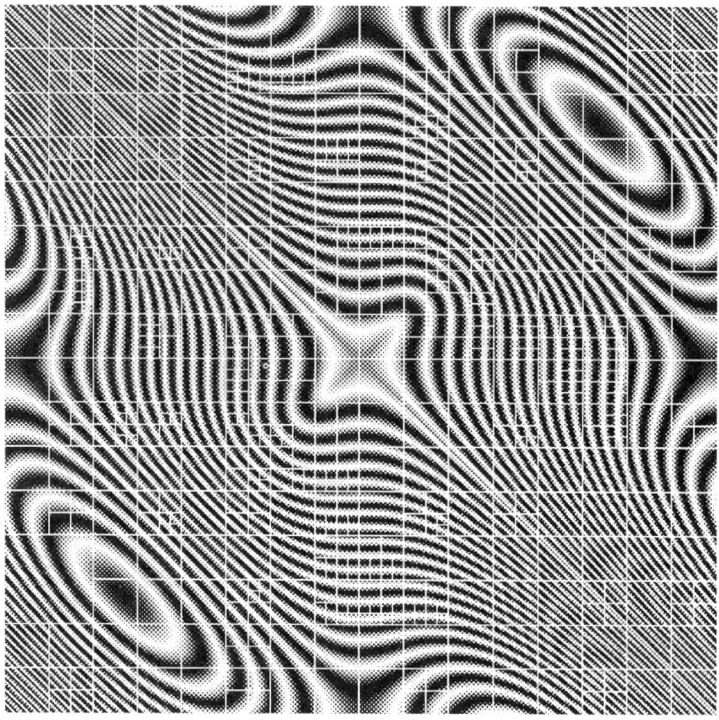

FIGURE 18. LCT window selection for a matrix compression.

Moreover, they form a basis for all functions f with finite square integral $\int_{-\infty}^{\infty} |f(x)|^2 \, dx < \infty$. This means that we can represent such a function as

$$f(x) = \sum_{j,k} \langle f, h_k^j \rangle h_k^j(x).$$

The coefficients $d_k^j = \langle f, h_k^j \rangle$ are called the Haar wavelet coefficients.

In order to facilitate the transition between the functions point of view (x varies continuously) and the discrete (sampled) numerical approach we choose to discretize a function on a given scale by defining its "sampled" values as being averages on that scale, i.e. for a fixed j we define

$$s_k^j = 2^{j/2} \int_0^{2^{-j}} f(x + 2^{-j}k)dx = \langle f, \chi_k^j \rangle$$

where

$$\chi(x) = \begin{cases} 1 & 0 < x \leq 1 \\ 0 & x \leq 0 \text{ or } 1 \leq x \end{cases}$$

is called a scaling function $\chi_k^j(x) = 2^{j/2}\chi(2^j x - k)$ (the function is normalized so that $\langle \chi_k^j, \chi_k^j \rangle = 1$). The number $2^{j/2}s_k^j$ is the average of f on the interval $[2^{-j}k, 2^{-j}(k+1)]$.

We observe that $d_k^{j-1} = \frac{1}{\sqrt{2}}(s_{2k}^j - s_{2k+1}^j)$ and that $s_k^{j-1} = \frac{1}{\sqrt{2}}\left(s_{2k}^j + s_{2k+1}^j\right)$, from which we deduce the following recursive algorithm for computing the Haar coefficients:

(6.1)
$$
\begin{array}{ccccccccc}
s^0 & \longrightarrow & s^1 & \longrightarrow & s^2 & \longrightarrow & s^3 & \cdots \\
 & \searrow & & \searrow & & \searrow & \\
 & & d^1 & & d^2 & & d^3 & \cdots
\end{array}
$$

Interpretation: s_k^j represents the time average of the signal f on the kth time interval of length 2^{-j}, d_k^j represents the variation of the average time signal on two consecutive intervals of length 2^{-j-1}.

7. Haar multiscale analysis.

We observe that for j fixed the functions χ_k^j are mutually orthogonal as k varies and that the map

$$f \to \sum_k \langle f, \chi_{jk} \rangle \chi_{j,k}(x) = P^j(f)$$

is the orthogonal projection on the space of functions which are constant on the intervals $(2^{-j}k, 2^{-j}(k+1))$ of length 2^{-j} (the sampling map). We will call this space V_j and observe that

$$V_j \supset V_{j+1} \supset V_{j+2}\ldots, \ \cup V_j = L^2 \text{ and } \cap V_j = \{0\}.$$

Moreover, if $f \in V_j$, then $f(2^k x) \in V_{j+k}$. We also observe that the orthogonal complement of V_j in V_{j+1} is spanned by the Haar wavelets. We will write

$$V_{j+1} = V_j \oplus W_j.$$

Exercise. Show that $P_{j+1}(f) - P_j(f) = \sum^k \langle f, h_k^j \rangle h_k^j$. We denote $P_{j+1} - P_j = Q_j$ or $P_{j+1} = P_j + Q_j$. We see that Q_j provides the detail needed to refine the sampling from the averages on scale j (resolution 2^{-j}) to the scale $j+1$ (resolution 2^{-j-1}). More generally,

$$P_N = P_0 + Q_0 + Q_1 + Q_2 \cdots + Q_{N-1}.$$

Here P_0 is the average signal on intervals of length 1, Q_0 adds the detail to obtain averages on intervals of length $\frac{1}{2}$, Q_1 adds refinements to intervals of length $\frac{1}{4}$, etc.

8. Walsh functions and a library of Haar-Walsh wavelet packets.

We now review the method for computing Haar coefficients: we restrict our attention to a sequence of eight samples $x_1, \ldots, x_8$ (which can be thought of as averages on intervals of length $\frac{1}{8}$ of a function defined on $[0, 1]$). The first computation involved

$$s_1 = \frac{1}{\sqrt{2}}(x_1 + x_2) = \frac{1}{\sqrt{2}}(x_1, x_2, \ldots, x_8) \cdot (1, 1, 0 \ldots 0)$$

$$d_1 = \frac{1}{\sqrt{2}}(x_1 - x_2) = \frac{1}{\sqrt{2}}(x_1, x_2, \ldots, x_8) \cdot (1, -1, 0, \ldots 0)$$

$$s_2 = \frac{1}{\sqrt{2}}(x_3 + x_4) = \frac{1}{\sqrt{2}}(x_1, x_2, \ldots, x_8) \cdot (0, 0, 1, 1, 0 \ldots 0)$$

$$d_2 = \frac{1}{\sqrt{2}}(x_3 - x_4) = \frac{1}{\sqrt{2}}(x_1, x_2, \ldots, x_8) \cdot (0, 0, 1, -1, 0 \ldots 0)$$

etc ...

Observe that the transformation from x to $\{s\}$ and $\{d\}$ consists of a string of rotations by $\frac{\pi}{4}$ of the vectors $(x_1, x_2)(x_3, x_4) \ldots$. Therefore, $x_1^2 + x_2^2 \cdots + x_8^2 = s_1^2 + s_2^2 + s_3^2 + s_4^2 + d_1^2 + d_2^2 + d_3^2 + d_4^2$, i.e. the total energy is conserved.

In the second stage we view $s_1 s_2 s_3 s_4$ as new samples (they are "averages" on intervals of length $\frac{1}{4}$) and repeat the procedure, computing sums of sums ss_1 ss_2 and differences of sums ds_1 ds_2.

It is natural to also view the differences $d_1 \ldots d_4$ (which measure the variation of the samples) as a new signal and perform the same transformations on them: sd_1 sd_2 corresponding to average variation, dd_1 dd_2 corresponding to change in variation, and continuing to fill in the rectangle, row by row, as in Figure 19. (This procedure will be interpreted later as subband coding).

x_1	x_2	x_3	x_4	x_5	x_6	x_7	x_8
s_1	s_2	s_3	s_4	d_1	d_2	d_3	d_4
ss_1	ss_2	ds_1	ds_2	sd_1	sd_2	dd_1	dd_2
sss_1	dss_1	sds_1	dds_1	ssd_1	dsd_1	sdd_1	ddd_1

FIGURE 19. A rectangle of wavelet packet coefficients.

We observe that each entry in this rectangular array of numbers represents an inner product of the original signal $(x_1, \ldots, x_8)$ with a multiple of a vector with entries ± 1 as described in the following diagrams:

For example, the entry dd_1 is obtained by taking

$$
\begin{aligned}
\frac{1}{\sqrt{2}}(d_1 - d_2) &= \frac{1}{2}(x_1, x_2, \ldots, x_8)(1, -1, 0, 0, \ldots 0) \\
&\quad - \frac{1}{2}(x_1, x_2, \ldots, x_8)(0, 0, 1, -1, \ldots 0) \\
&= \frac{1}{2}(x_1, x_2, \ldots, x_8)(1, -1, -1, 1, 0, 0, 0, 0)
\end{aligned}
$$

which is the pattern corresponding to the first box in the 4th block on level 2 in Figure 19 (the signal is on level 0).

The patterns (or vectors) generated in the preceding pages can be combined in different ways to construct orthogonal bases of eight dimensional space.

The last eight patterns on level 3 represent the well known Walsh pattern functions (providing a square wave Fourier analysis). Clearly the transform mapping the original sequence into the entries of the bottom row is orthogonal (since it was obtained by a succession of orthogonal transformations). Therefore the different patterns which are the columns of this transformation are orthogonal. The basis corresponding to a fixed row provides a windowed Walsh transform.

The discrete Haar wavelet basis is obtained by choosing the second block in each row and the first and second entry on the last row.

It is easy to see that any collection of blocks in the rectangle with the property that their shadow intervals form a disjoint cover of the full range provides a collection of patterns forming a basis. As can be seen on the following example diagram, we use the entries on the bottom level (3) to recover the entries on the "parent" box above these entries. We then use the entries on level 2 to recover all entries on level in the parent box. We now have a full set of entries on level one enabling us to recover the original signal. Since all transformations were orthogonal we must have that the collection of vectors corresponding to this choice of patterns is an orthogonal basis of $\mathbf{R}^8$.

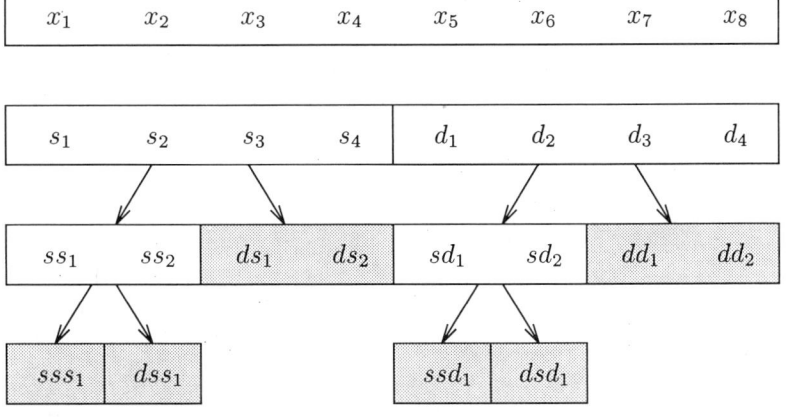

FIGURE 20.

General wavelet packets. We'll use the notation and terminology of Daubechies (1988), whose results we shall assume.

We are given an exact quadrature mirror filter $h(n)$ satisfying the conditions of Theorem 3.6 in Daubechies (1988), p. 964, i.e.

$$\sum h(n-2k)h(n-2\ell) = \delta_{k,\ell} \, , \quad \sum h(n) = \sqrt{2}.$$

We let $g_k = \overline{h_{-k+1}}(-1)^k$ and define the operations F_i on $\ell^2(\mathbf{Z})$ into "$\ell^2(2\mathbf{Z})$"

(8.1)
$$F_0\{s_k\}(i) = 2\sum s_k h_{k-2i}$$

$$F_1\{s_k\}(i) = 2\sum s_k g_{k-2i}.$$

The map $\mathbf{F} : s_k \mapsto F_0(s_k) \oplus F_1(s_k) \in \ell^2(2\mathbf{Z}) \oplus \ell^2(2\mathbf{Z})$ is orthogonal and

(8.2)
$$F_0^* F_0 + F_1^* F_1 = I$$

We now define the following sequence of functions.

(8.3)
$$\begin{cases} W_{2n}(x) = \sqrt{2}\sum h_k W_n(2x-k) \\ W_{2n+1}(x) = \sqrt{2}\sum g_k W_n(2x-k). \end{cases}$$

Clearly the function $W_0(x)$ can be identified with the function φ in Daubechies (1988) and W_1 with the function ψ.

Let us define $m_0(\xi) = \frac{1}{\sqrt{2}}\sum h_k e^{-ik\xi}$ and

$$m_1(\xi) = -e^{-i\xi}\overline{m_0(\xi+\pi)} = \frac{1}{\sqrt{2}}\sum g_k e^{-ik\xi}$$

All of the functions W_n have a fixed scale, but we observe that mixed-scale decompositions of L^2 are also possible. This allows us to refine the decomposition L^2 by scales as embodied in the following:

THEOREM 8.1. *For every partition P of the nonnegative integers into sets of the form $I_{kn} = \{2^k n, \ldots, 2^k(n+1) - 1\}$, the collection of functions $\{2^{k/2}W_n(2^k t - \ell) : I_{kn} \in P, \ \ell \in \mathbf{Z}\}$ is an orthonormal basis for $L^2(\mathbf{R})$.*

Remark. We may also think of I_{nk} as the dyadic subinterval $[2^{-k}n, 2^{-k}(n+1)[$ of $[0,1[$. Such an indexing convention gives a faithful correspondence between disjoint dyadic decompositions of the frequency line and orthonormal wavelet packet subsets of L^2.

DEFINITION 8.2. *A wavelet packet basis of $L^2(\mathbf{R})$ is any orthonormal basis selected from among the functions $2^{k/2}W_n(2^k t - \ell)$.*

Beside the Walsh-type basis, examples of wavelet packet bases include the *wavelet basis* and the *subband basis*.

A useful picture of the tree of wavelet packet coefficients is that of a rectangle of dyadic blocks. The row number within the rectangle indexes the scale of the wavelet packets listed therein. The column number indexes both the frequency and position parameters. We may choose to group the wavelet packets either by position or by frequency. Grouping by position fills each row of the rectangle with adjacent windowed spectral transforms, analogous to windowed FFT, with the window size determined by the row number and the window position corresponding to the location of the group. The frequency parameter increases within the group.

We will group the coefficients by frequency, since that gives a more efficient implementation, and since the transformation to the other form is obvious. The boxes of coefficients in the rectangle correspond to the decomposition of $\delta^L \Omega_0$ into the subspaces $\delta^k \Omega_n$, for $0 \leq k \leq L$ and $0 \leq n < 2^{L-k}$. The top box corresponds to $\delta^L \Omega_0$, the bottom boxes correspond to Ω_n, for $0 \leq n < 2^L$, and box n on level k (counting the bottom as level 0) corresponds to subspace $\delta^k \Omega_n$.

For definiteness we return to the Haar example. Consider a function defined at 8 points $\{x_1, \ldots, x_8\}$, i.e., a vector in $\mathbf{R}^8$. Then the (periodized) wavelet packet coefficients of this function can be represented as in Figure 19 above. In Figure 19, each row is computed from the row above it by one application of either F_0 or F_1, which we think of as "summing" (s) or "differencing" (d) operations, respectively. Thus, for example, the subblock $\{ss_1, ss_2\}$ comes from the application of F_0 to $\{s_1, s_2, s_3, s_4\}$, while $\{ds_1, ds_2\}$ comes similarly from F_1. The two descendent s and d subblocks on row $n + 1$ are determined by their mutual parent on row n, which conversely is determined by them through the adjoint anticonvolutions F_0^* and F_1^*. The process is depicted in this exploded view:

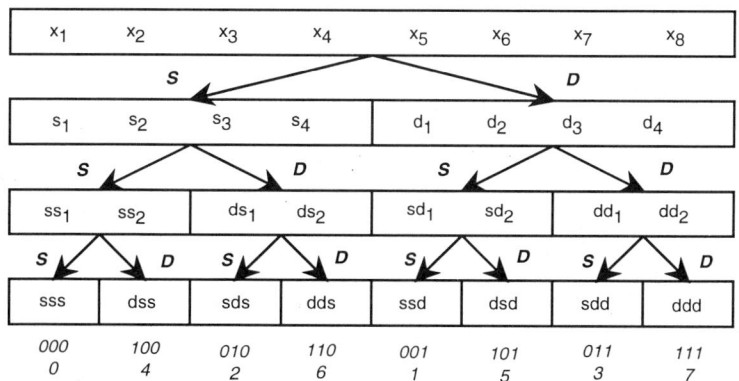

FIGURE 21. Naturally ordered wavelet packets on $\mathbf{R}^8$.

In the simplest case, where we use the Haar filters $h = \{\frac{1}{\sqrt{2}}, \frac{1}{\sqrt{2}}\}$ and $g = \{\frac{1}{\sqrt{2}}, -\frac{1}{\sqrt{2}}\}$, we have in particular $ss_1 = \frac{1}{\sqrt{2}}(s_1 + s_2)$, $ss_2 = \frac{1}{\sqrt{2}}(s_3 + s_4)$, $ds_1 = \frac{1}{\sqrt{2}}(s_1 - s_2)$, and $ds_2 = \frac{1}{\sqrt{2}}(s_3 - s_4)$. For $N = 8$, we can draw the functions which correspond to the entries in the rectangle. These are displayed in the figures below.

Smallest scale = level 1;   Intermediate scale = level 2;   Largest scale = level 3.

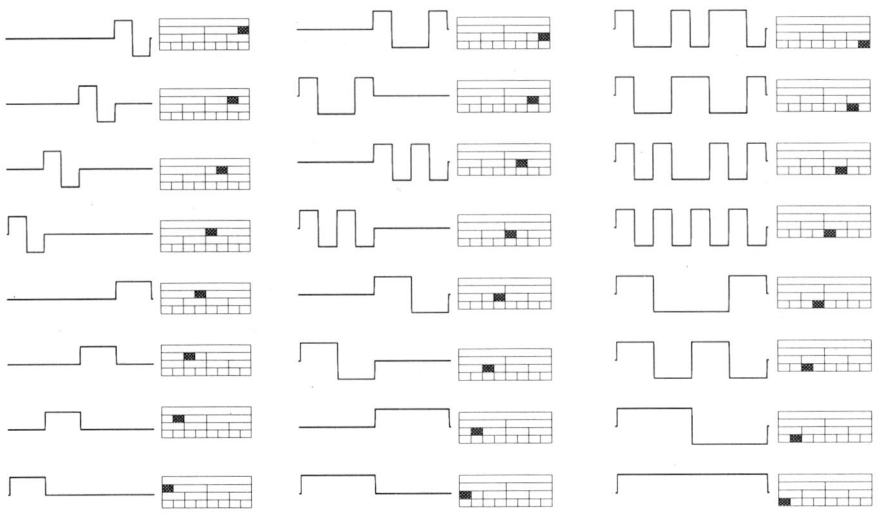

FIGURE 22. Haar wavelet packets on $\mathbf{R}^8$.

The algorithm produces Haar wavelet packets in the "Paley" or "natural" order. The algorithm may be easily modified to produce "sequency" ordered wavelet packets: what is needed is to exchange F_0 and F_1 whenever the parent's sequency is odd. The diagram below depicts the permuted wavelet packet transform on an 8 point signal, which produces Walsh functions in sequency order if we use the Haar filters above for F_0 and F_1. Sequency has a strict definition only for Walsh functions, where it is the number of zero-crossing of a function which takes only the values 1 and -1. The nth Shannon wavelet packet, in sequency order, is band-limited to the intervals $\pm[n, n+1[$. If we define the appropriate notion of "main frequency" in the intermediate case of smooth, compactly supported wavelet packets, we see that main frequency increases monotonically with sequency order. Paley order can also be obtained from sequency order by the Gray code permutation. Note that our restriction to signals with 8 samples only and to the Haar filters was purely for illustration purposes; the same ideas apply to other wavelet filters and longer signals.

9. Multidimensional wavelet packets transforms.

The quadrature mirror filter algorithm extends to multidimensional signals by separation of variables. Let F_0 and F_1 be a pair of QMF's for $\mathbf{R}$; then $F_0 \otimes F_0$, $F_0 \otimes F_1$, $F_1 \otimes F_0$, and $F_1 \otimes F_1$ are a family of 4 filters for $\mathbf{R}^2$. In general, for d-dimensional signals, we will use a 2^d member family of filters $\otimes_{i=1}^{d} F_{\varepsilon_i}$, where $\varepsilon_i \in \{0,1\}$ for $i = 1, \ldots, d$. The wavelet packets produced by iterating these filters are products of one-dimensional wavelet packets $W_n(x) = \prod_{i=1}^{d} W_{n_i}(x_i)$,

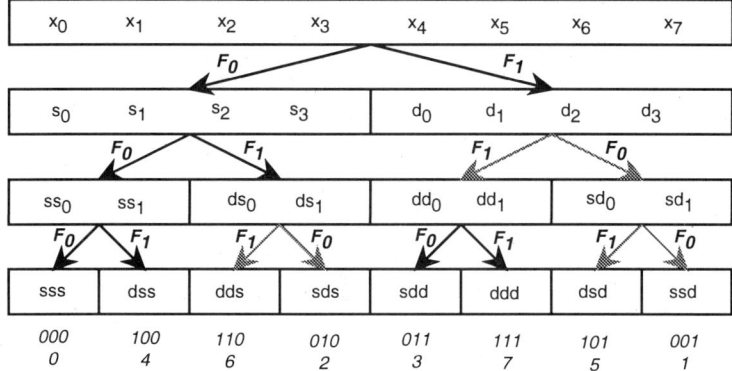

| x_0 | x_1 | x_2 | x_3 | x_4 | x_5 | x_6 | x_7 |

F_0 F_1

| s_0 | s_1 | s_2 | s_3 | d_0 | d_1 | d_2 | d_3 |

F_0 F_1 F_1 F_0

| ss_0 | ss_1 | ds_0 | ds_1 | dd_0 | dd_1 | sd_0 | sd_1 |

F_0 F_1 F_1 F_0 F_0 F_1 F_1 F_0

| sss | dss | dds | sds | sdd | ddd | dsd | ssd |

| 000 | 100 | 110 | 010 | 011 | 111 | 101 | 001 |
| 0 | 4 | 6 | 2 | 3 | 7 | 5 | 1 |

FIGURE 23. Sequency ordered wavelet packets on $\mathbf{R}^8$.

together with their isotropic dilations and translations to arbitrary lattice points. Inner products with these multidimensional wavelet packets are computed from averages at the smallest scale, just as in the one-dimensional case. The coefficients may be organized into a stack of d-dimensional intervals, and there is a result analogous to Theorem 8.1 which characterizes orthogonal basis subsets. We can restrict our attention to the case $d = 2$. The diagram below depicts a decomposition of the unit square into a 3-level wavelet packet tree:

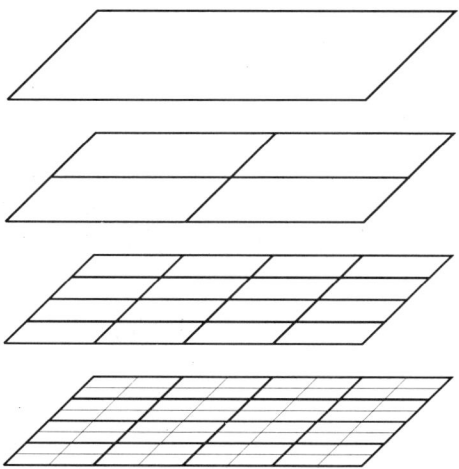

FIGURE 24. 2-dimensional wavelet decomposition to level 3.

Abusing notation, we can write $FF_0 = F_0 \otimes F_0$, $FF_1 = F_0 \otimes F_1$, $FF_2 = F_1 \otimes F_0$, and $FF_3 = F_1 \otimes F_1$. Note that the numbering corresponds to $\varepsilon_1 \varepsilon_2$ in binary. Each 2-dimensional filter decimates by 2 in both in the x and y directions, so it reduces the number of coefficients by 4. The coefficients in the various boxes of this picture are calculated by the application of the filters as labelled below.

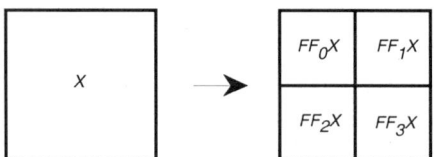

FIGURE 25. 2-dimensional wavelet packet coefficients.

By applying the filter convolutions recursively, we obtain a homogeneous tree-structured decomposition organized by frequency.

10. Denoising and coherent structure extraction.

As an application combining these ideas we now describe an algorithm for denoising or coherent structure extraction

We choose to view a signal f as being noisy or incoherent relative to the basis if it does not correlate well with the waveforms of the basis ω_i i.e. if its entropy is of the same order of magnitude as $\log_2 N - \delta$, giving a poor compression rate $2^{-\delta}$. This leads us to the following method.

Start with a signal f of length N, find the best basis in each library and select among them the basis minimizing $\varepsilon(f)$. Reorder the coefficients α_i in decreasing order $\alpha_1 \geq \alpha_2 \geq \cdots \alpha_{N_0} > 0$ where α_i for $i \geq N_0$ are all below a precision threshold (say 0.1% of energy). Then decompose

$$f = c_M + r_M \text{ where } c_M = \sum_1^M \alpha_i \omega_i, \quad r_M = \sum_{M+1}^{N_0} \alpha_i \omega_i.$$

We will say that c_M is coherent and r_M is incoherent if $c(r_M) \geq \tau_0$, where $c(v)$ denotes the compression ratio of the decomposition of v, i.e. the number of bits needed to encode the coefficients in the decomposition (here the α_i, $i = M + 1, \ldots, N_0$ if $v = r_M$) versus the number of bits in the "raw data" v. The threshold τ_0 is chosen to determine if the compression of r_M using ω_i is unacceptably bad. We proceed by testing $r_1 r_2 \ldots$ until we reach M for which

$$c(r_M) \geq \tau_0 > 0 , \quad 0 < \tau_0 < 1$$

or

$$\sum_{i=M_0+1}^{N_0} \frac{\alpha_i^2}{\|r_M\|^2} \log_2 \left(\frac{\|r_M\|^2}{\alpha_i^2} \right) \geq \log_2(N_0 - M_0 + 1) - \log_2 \tau_0.$$

Following the procedure proposed by Mallat we can now consider r_M as a new signal for which we repeat the decomposition, i.e. pick a best basis (different from the best basis at the preceding step) and decompose

$$r_M = c'_{M_1} + r'_{M_1}$$

iterating a fixed number of times or stopping whenever no new coherent part is obtained.

In Figure 26 an original underwater sound signal is peeled into layers as described above.

REFERENCES

1. B. Alpert and V. Rokhlin, "A Fast Algorithm for the Evaluation of Legendre Expansions", *SIAM J. of Scientific and Statistical Computing* **14** (1993) 159–189.
2. J. Carrier, L. Greengard and V. Rokhlin, *A Fast Adaptive Multipole Algorithm for Particle Simulations,* Yale University Technical Report, YALEU/DCS/RR-496 (1986), SIAM Journal of Scientific and Statistical Computing, to appear **9(4)** (1988).
3. R. Coifman and Y. Meyer, *Non-linear Harmonic Analysis, Operator Theory and P.D.E.,* Annals of Math Studies, Princeton, 1986, ed. E. Stein.
4. R. Coifman and Y. Meyer, "Remarques sur l'analyse de Fourier à fenêtre", série I, *C. R. Acad. Sci. Paris* **312** (1991) 259–261.
5. R. Coifman, Y. Meyer, Steven Quake, M. Victor Wickerhauser, *Signal Processing and Compression With Wavelet Packets,* Proceedings of the International Conference on Wavelets, Toulouse, 1992, Y. Meyer and S. Roques (eds.), to be published by Springer.
6. R. Coifman, Y. Meyer and V. Wickerhauser, "Wavelet Analysis and Signal Processing", pp. 453–470 in *Wavelets and their applications,* M. B. Ruskai et al. (eds.), Jones and Bartlett (Boston) 1992.
7. R. Coifman and M. V. Wickerhauser, "Entropy-Based Algorithms for Best Basis Selection", *IEEE Trans. Inf. Th.* **38** (1992) 713–718.
8. I. Daubechies, "Orthonormal bases of compactly supported wavelets", *Communications on Pure and Applied Mathematics* **XLI** (1988), 909–996.
9. I. Daubechies, S. Jaffard and J.L. Journé, "A simple Wilson orthonormal basis with exponential decay", *SIAM J. Math. Anal.* **22** (1991) 554–572.
10. L. Greengard and V. Rokhlin, "A Fast Algorithm for Particle Simulations", *Journal of Computational Physics,* 73(1) **325** (1987).
11. S. Mallat, *Review of Multifrequency Channel Decomposition of Images and Wavelet Models,* Technical Report 412, Robotics Report 178, NYU (1988).
12. H. Malvar, "Lapped transforms for efficient transform/subband coding", *IEEE Trans. ASSP* **38** (1990) 969–978.
13. Y. Meyer, *Principe d'incertitude, bases hilbertiennes et algèbres d'opérateurs,* Séminaire Bourbaki, 1985-86, 662, Astérisque (Société Mathématique de France).
14. Y. Meyer, *Wavelets and Operators,* Analysis at Urbana vol.1 edited by E. Berkson, N.T. Peck and J. Uhl, London Math. Society, Lecture Notes Series 137, 1989.
15. S.T. O'Donnel and V. Rokhlin, *A Fast Algorithm for the Numerical Evaluation of Conformal Mappings,* Yale University Technical Report, YALEU/DCS/RR-554 (1987), SIAM Journal of Scientific and Statistical Computing, submitted.
16. J.O. Stromberg, *A Modified Haar System and Higher Order Spline Systems,* Conference in harmonic analysis in honor of Antoni Zygmund, Wadworth math.series, edited by W. Beckner et al. **II**, 475–493.

MATHEMATICS DEPARTMENT, YALE UNIVERSITY, NEW HAVEN, CT 06520
E-mail: coifman_ronald@math.yale.edu

DEPARTMENT OF MATHEMATICS, WASHINGTON UNIVERSITY, ST. LOUIS, MISSOURI 63130
E-mail: victor@math.wustl.edu

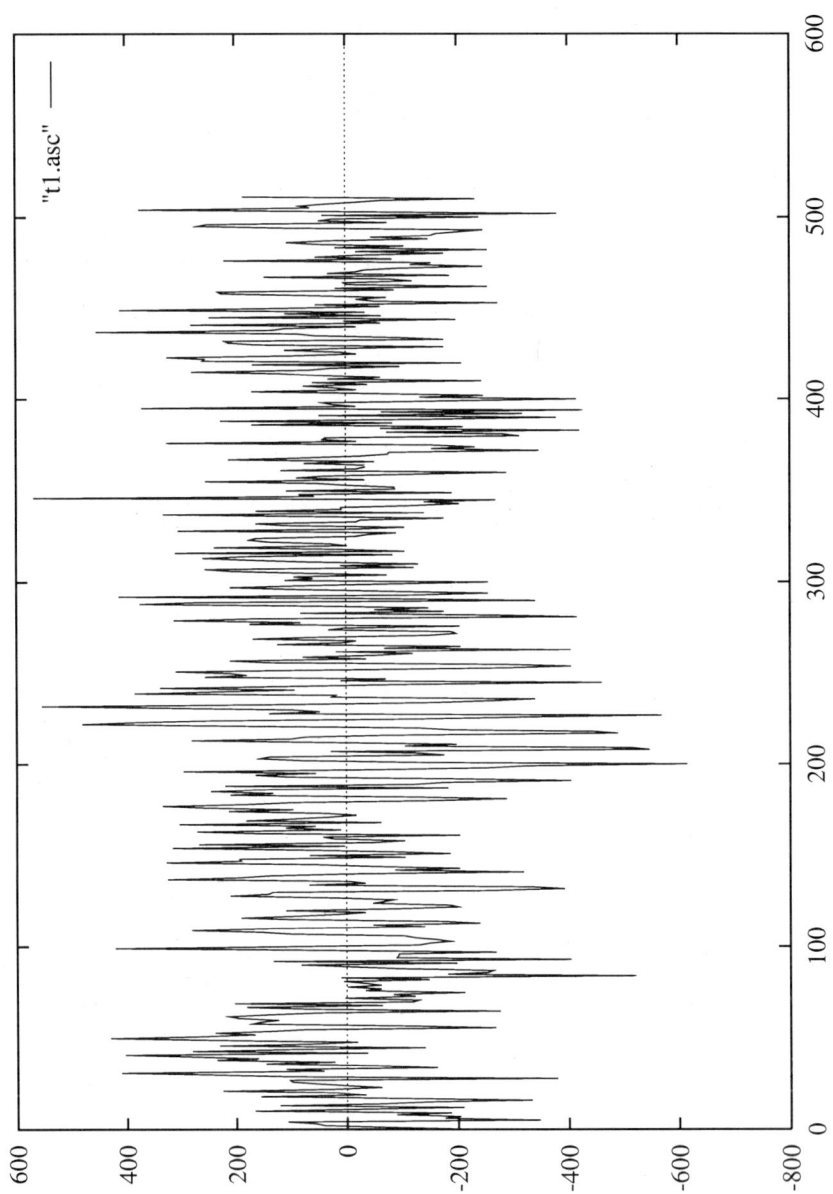

FIGURE 26. (a) An underwater sound signal.

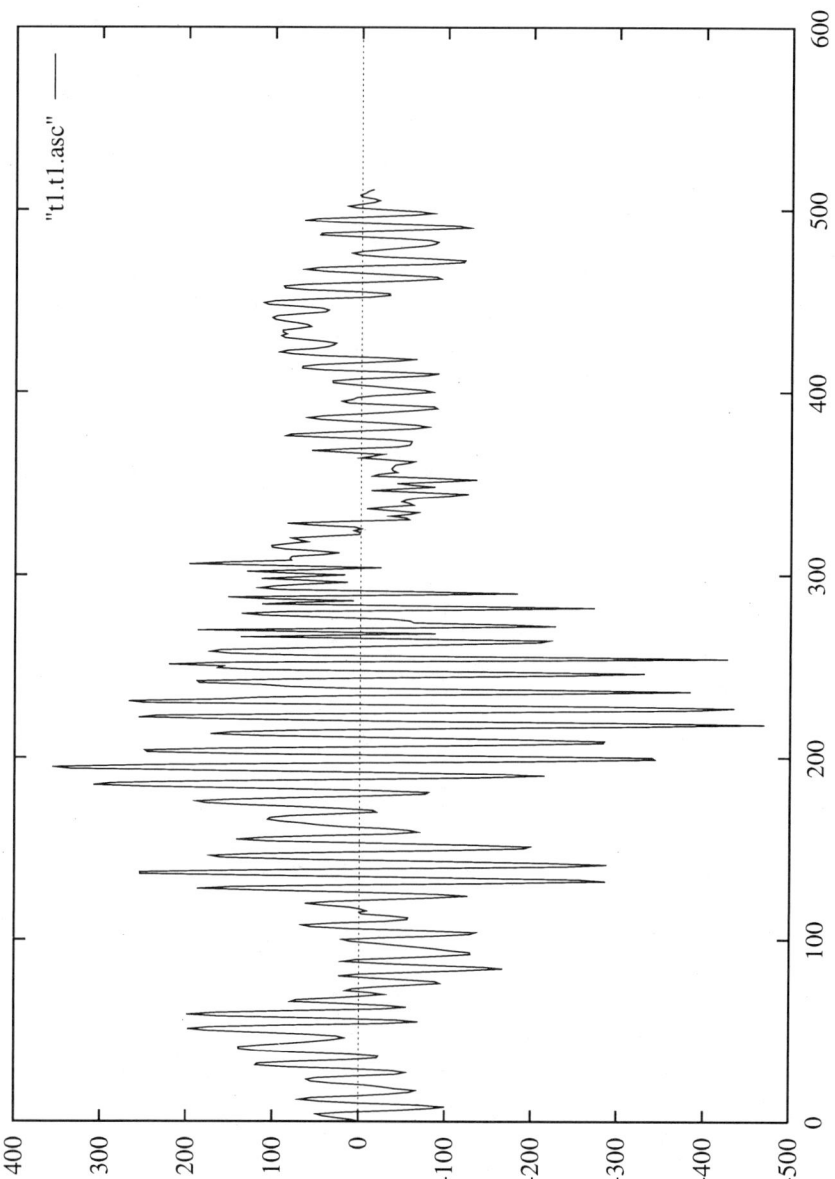

FIGURE 26. (b) The first coherent component of the underwater sound signal.

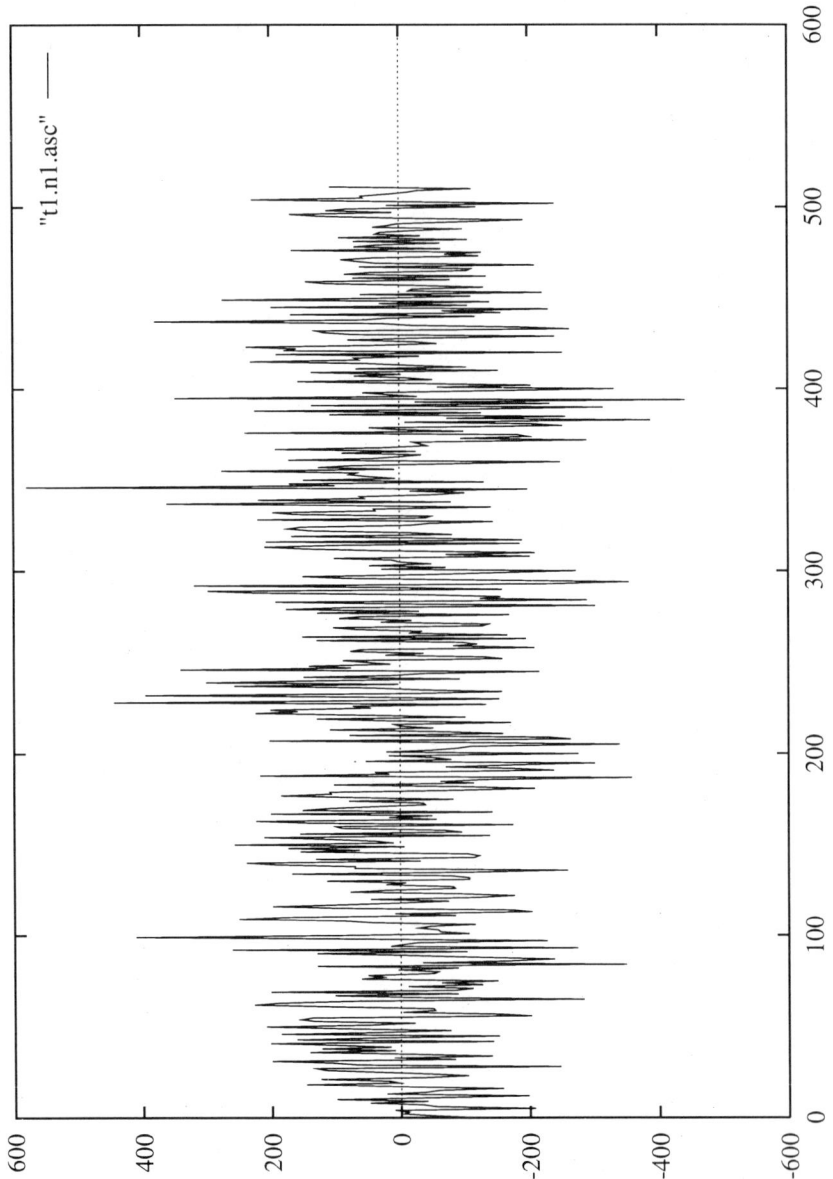

FIGURE 26. (c) The first noisy residue of the underwater sound signal.

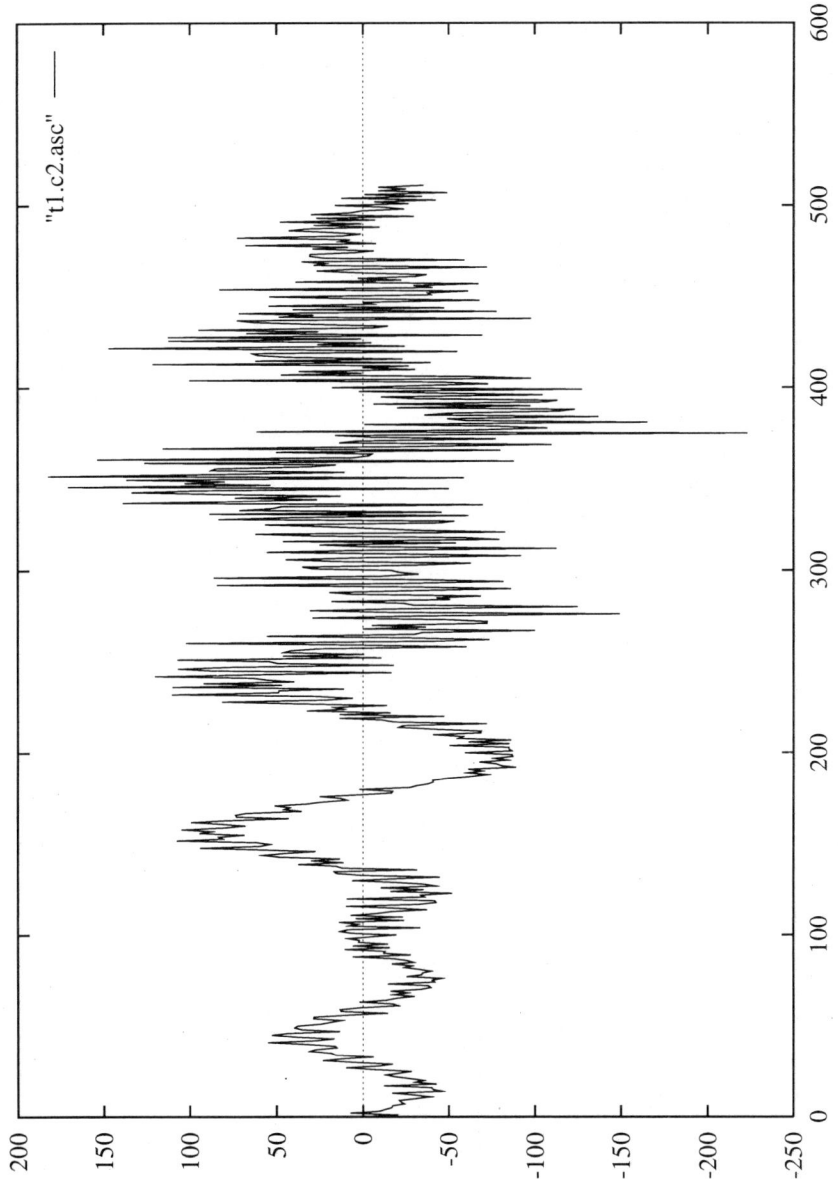

FIGURE 26. (d) The coherent part of the preceding noisy part.

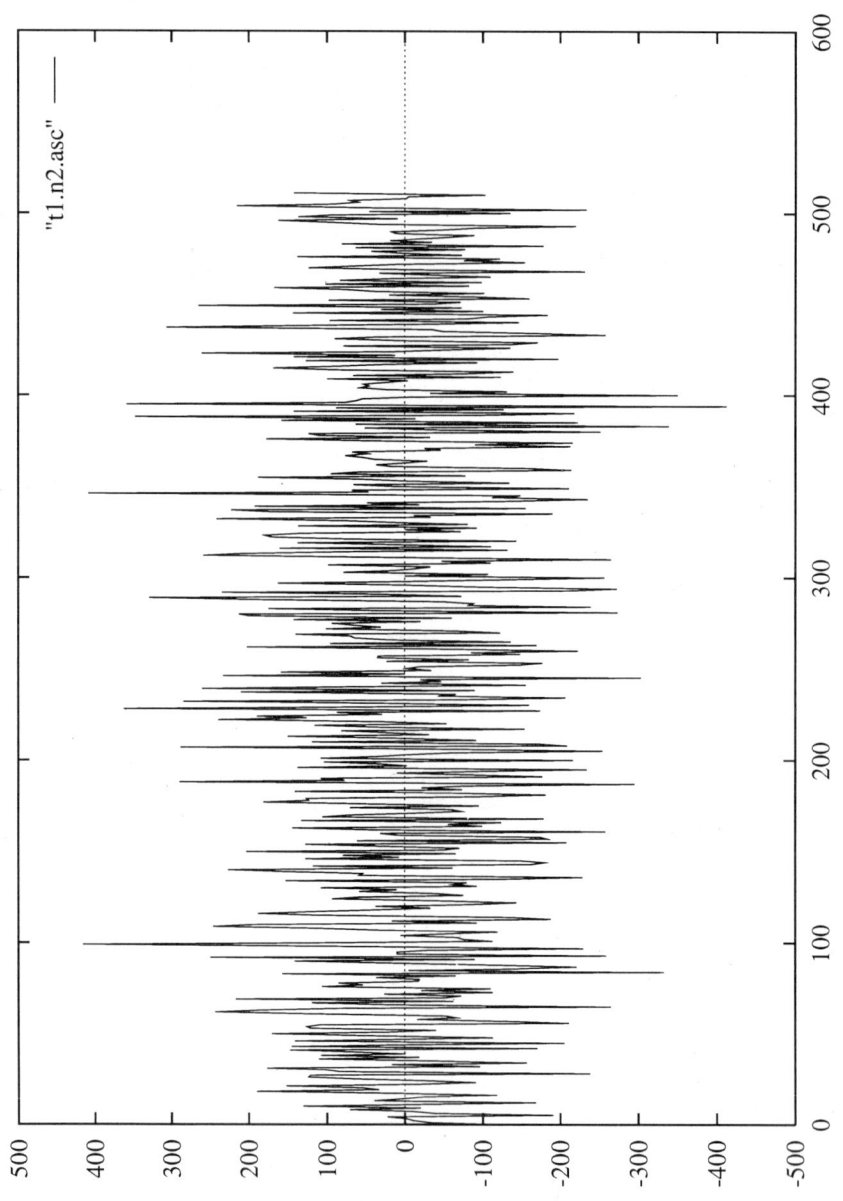

FIGURE 26. (e) The second noisy component.

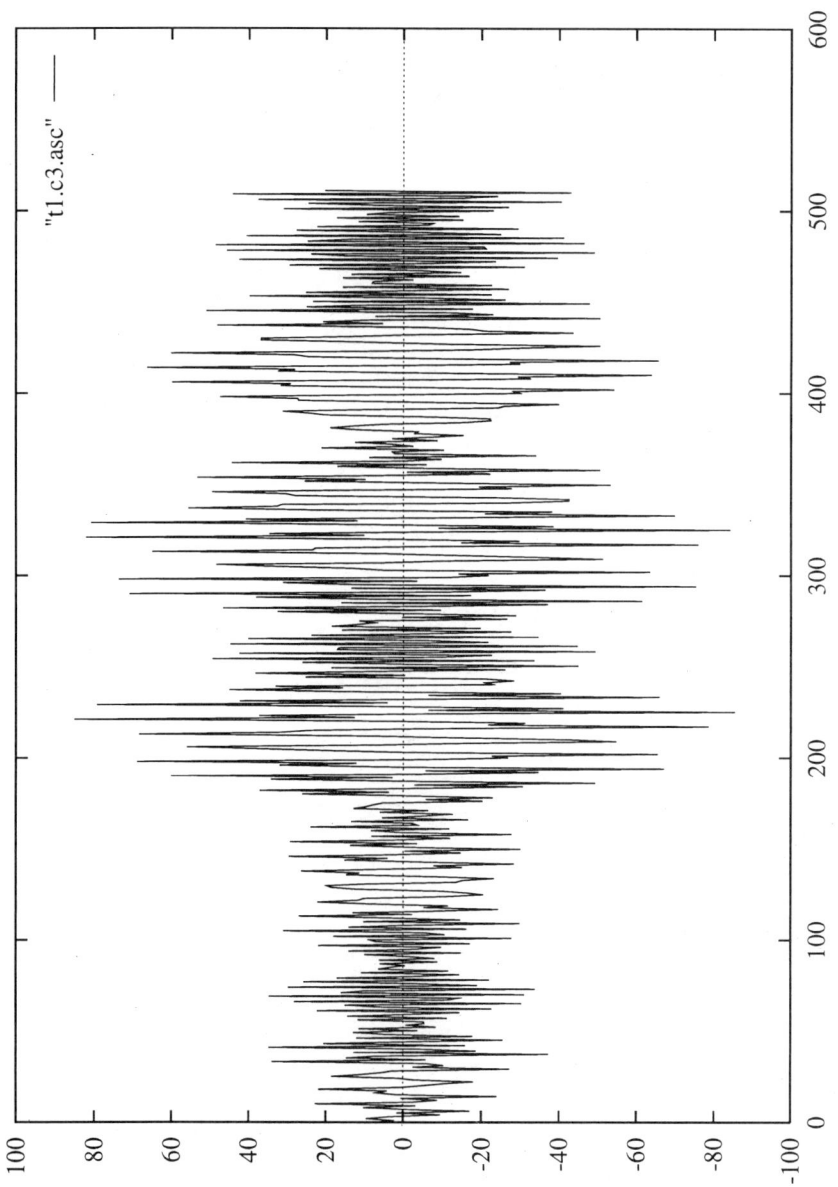

FIGURE 26. (f) The third coherent component.

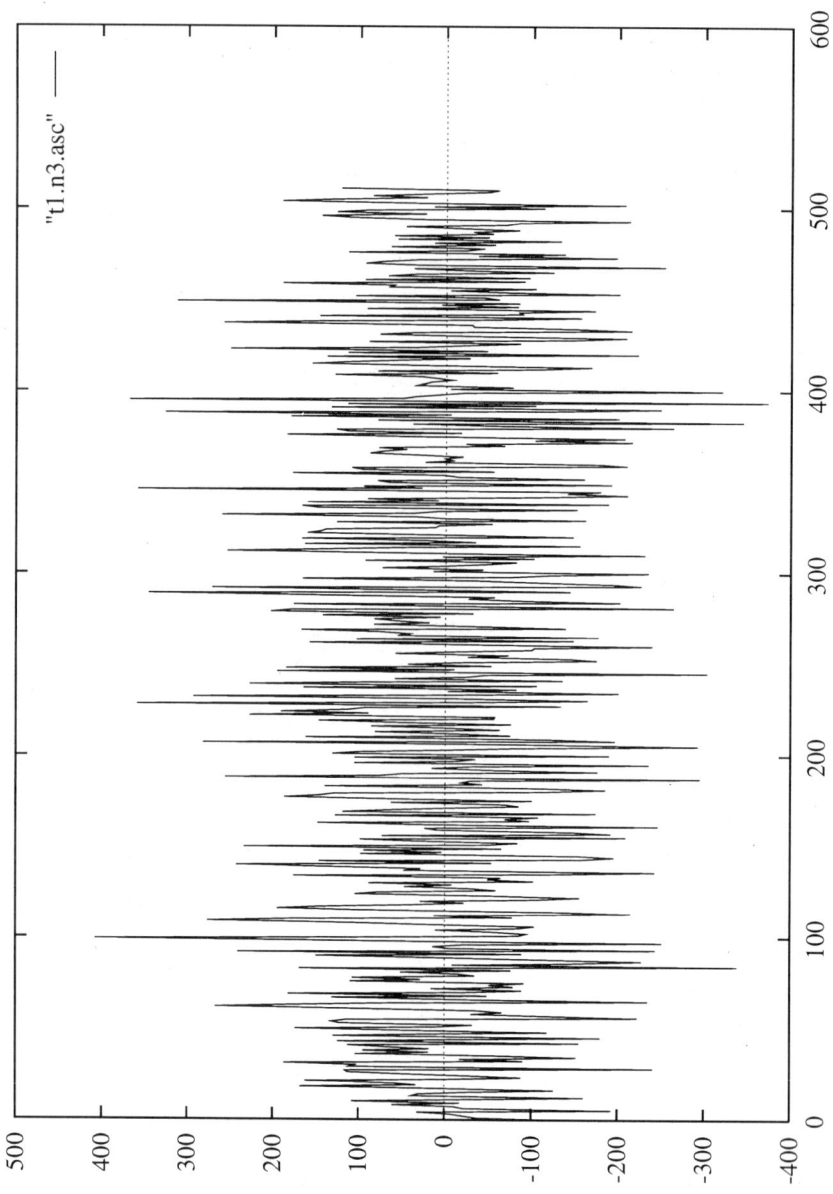

FIGURE 26. (g) The third noisy component.

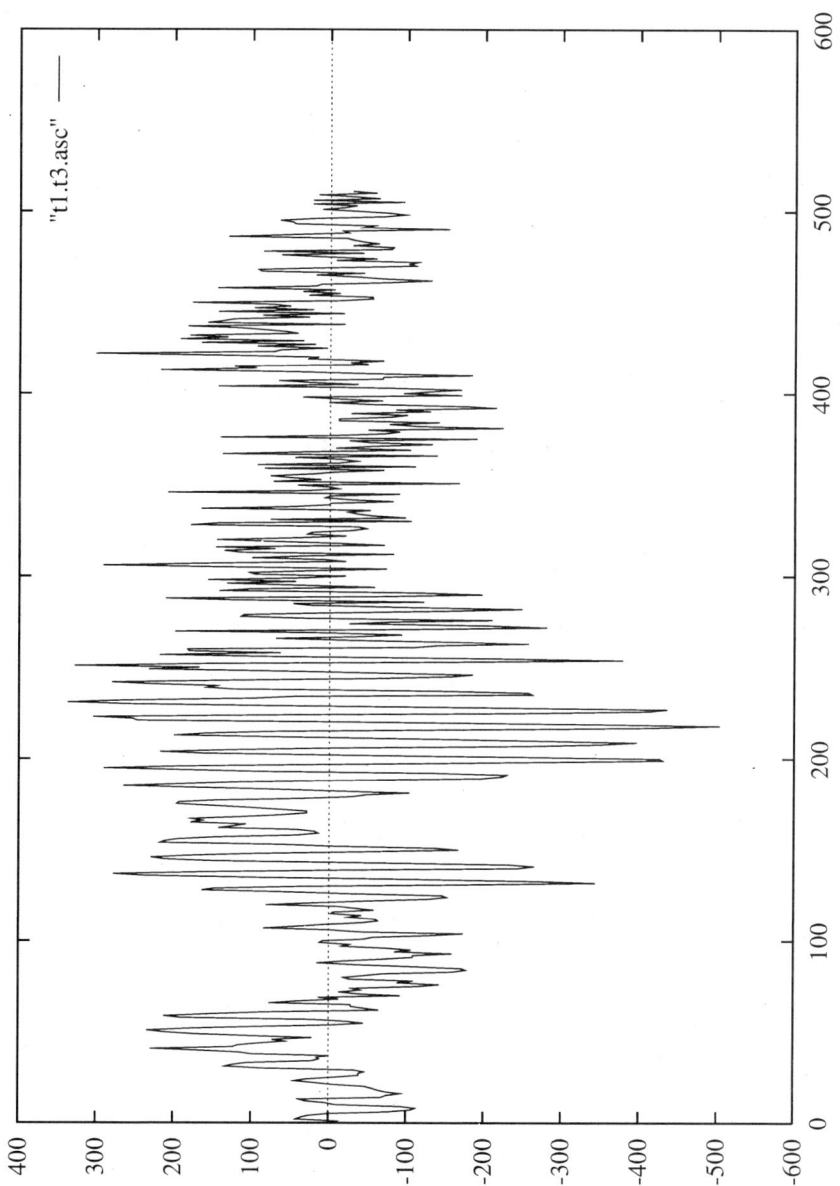

FIGURE 26. (h) The sum of the coherent parts.

Proceedings of Symposia in Applied Mathematics
Volume **47**, 1993

Best-adapted Wavelet Packet Bases

MLADEN VICTOR WICKERHAUSER

ABSTRACT. This paper is a review of the construction of orthogonal wavelet packets, using the quadrature mirror filter algorithm slightly generalized to the case of $p \geq 2$ wavelets and scaling functions.

1. Introduction.

We begin with a classical reproducing formula for functions $f \in H^2$, the Hardy space of square-integrable functions whose Fourier transforms vanish on the negative half-line:

$$\text{If} \quad c = 2\pi \int_0^\infty \frac{|\hat{\psi}(\xi)|^2}{|\xi|}\, d\xi < \infty \quad \text{and} \quad T_f(a,b) = \int_{\mathbf{R}} f(x)\overline{\psi(ax+b)}\, dx,$$

$$\text{then} \quad f(x) = \frac{1}{c} \iint_{\mathbf{R} \times \mathbf{R}^+} Wf(a,b)\psi(ax+b)\, da\, db$$

This formula was studied by Calderón in the 60's and revived by Grossmann and Morlet in their 1984 paper [**5**]. A function ψ satisfying the *admissibility condition* $c < \infty$ is called a "wavelet," and the map $f \mapsto T_f$ is called the (continuous) wavelet transform. The discrete (dyadic) wavelet transform transform is the restriction $f \mapsto \{T_f(2^j, k),\ j, k \in \mathbf{Z}\}$. A well-established sampling theory [**3**], [**10**] exists which provides necessary and sufficient conditions on ψ for the discrete transform to be bijective. There are compactly-supported functions ψ with any given degree of smoothness for which the discrete wavelet transform is orthogonal [**2**],[**8**]; there is also an orthonormal basis of C^∞ wavelets with decay faster than any inverse polynomial [**9**].

The "fast" discrete wavelet transform computes $\{T_f(2^j, k),\ j, k \in \mathbf{Z}\}$ by iterating a pair of operators called quadrature mirror filters (QMFs). Suppose that

1991 *Mathematics Subject Classification.* Primary 33C45, 42C10, 94A11.
Research supported in part by AFOSR Grant F49620-92-J-0106.

H, G are bounded linear operators defined by
(QMF)

$$Hz_n = \sum_k h_k z_{2n-k}, \quad Gz_n = \sum_k g_k z_{2n-k} \quad \text{from } \ell^2(\mathbf{Z}) \text{ to } \ell^2(\mathbf{Z});$$

$$Hf(x) = \sum_k h_k f(2x-k), \quad Gf(x) = \sum_k g_k f(2x-k) \quad \text{from } L^2(\mathbf{R}) \text{ to } L^2(\mathbf{R}).$$

They are QMFs if $HG^* = 0$, $HH^* = GG^* = I$, and $H^*H \oplus G^*G = I$. It is conventional to assume that $H\mathbf{1} = \sqrt{2}\mathbf{1}$ (where $\mathbf{1}$ is the sequence of 1's) and $G\mathbf{1} = 0$, and to call H and G the *low-pass* and *high-pass* filters, respectively. The fast discrete wavelet transform computes coefficients by the "pyramid scheme" depicted in the diagram below:

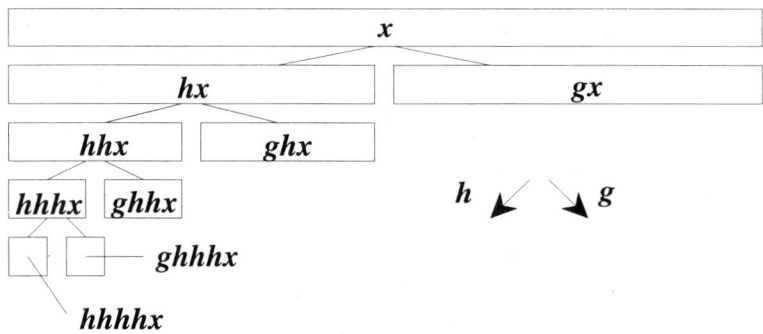

FIGURE 1. Pyramid scheme for the "fast" discrete wavelet transform.

The functions underlying the expansion are called "wavelets" and "scaling" functions or "mother" and "father" functions respectively. If $\{h_k\}$ and $\{g_k\}$ are finite sequences, then we have:

LEMMA 1.1. There is a unique function $\phi \in L^2 \cap L^1$ of compact support solving $H\phi = \phi$, $\int \phi = 1$.

LEMMA 1.2. The function $\psi = G\phi \in L^2 \cap L^1$ is a wavelet with compact support.

Meyer and Mallat introduced the *multiresolution analysis* of $L^2(\mathbf{R})$ (or MRA); it is a sequence of subspaces $\{V_j : j \in \mathbf{Z}\}$ based on a single function ϕ, with $V_j \stackrel{\text{def}}{=} \text{span}\{\phi(2^j x - k) : k \in \mathbf{Z}\}$ satisfying $j < j' \implies V_j \subset V_{j'}$, $\overline{\bigcup_j V_j} = L^2$, and $\bigcap_j V_j = 0$. Given an MRA, we may put $W_j = V_j^\perp \cap V_{j+1}$ to get $L^2 = \overline{\bigoplus_j W_j}$. If ϕ is the function in Lemma 1, then $\{V_j\}$ is an MRA and $\psi = G\phi$ is a wavelet. Conversely, it is known [**6**] that if the function ψ has compact support and the set $\{2^{j/2}\psi(2^j x - k) : j, k \in \mathbf{Z}\}$ is an orthonormal basis for L^2, then it comes from an MRA with a compactly supported ϕ.

There is a fast numerical functional calculus based on the compactly-supported orthonormal wavelet transform. It is implemented by defining orthogonal projections $P_j : f \twoheadrightarrow V_j$ and $Q_j : f \twoheadrightarrow W_j$; these are approximated either by sampling $P_j f(k) = \int f(x)\phi(2^j x - k)\,dx \approx 2^{-j} f(2^{-j}k)$ or by using some higher-order numerical quadrature formula. Thereafter, all computations within the MRA are performed by iterating the low complexity maps H and G on just those coefficients (from $\{P_j f(k) : k \in \mathbf{Z}\}$) which are larger than some threshold depending upon the numerical precision of the calculation.

Coifman, Meyer and Wickerhauser [1] have described a large "library" of orthonormal bases generalizing the wavelet basis. These bases consist of "wavelet packets" which are superpositions of wavelets and are described efficiently by short sequences of H and G. Wavelet packets come with independent frequency, position and duration parameters and can be used to build individually-adapted orthonormal bases for oscillatory functions and operator kernels.

By using extra filters, it is possible to introduce fast wavelet packet transformations which decimate by arbitrary numbers. Such transformations generalize algorithms which decimate by 2. The method produces new libraries of orthonormal basis vectors. We will describe the *best basis* algorithm for selecting a most efficient representation from this library. The extra generality of p filters is not expensive and in fact may clarify certain points. We will prove that the best-basis algorithm has complexity $O(N \log_p N)$ for a sequence of length N. We will also discuss some of the analytic properties and applications of such representations.

2. Aperiodic filters and bases in l^2.

Consider first the construction of bases on l^2. Let p be a positive integer and introduce p absolutely summable sequences $f_0, \ldots, f_{p-1}$ satisfying the properties:
  (i) For some $\epsilon > 0$, $\sum_m |f_i(m)||m|^\epsilon < \infty$,
  (ii) $\sum_m f_0(pm + i) = 1/\sqrt{p}$, for $i = 0, 1, \ldots, p - 1$, and
  (iii) $\sum_m f_i(m)f_j(m + kp) = \delta_{i-j}\delta_k$,   where δ is the Kronecker symbol.
To these sequences are associated p convolution operators $F_0, \ldots, F_{p-1}$ and their adjoints $F_0^*, \ldots, F_{p-1}^*$ defined by

$$F_i : l^2 \to l^2, \quad F_i v(k) = \sum_m f_i(m + pk)v(m),$$

$$F_i^* : l^2 \to l^2, \quad F_i^* v(m) = \sum_k \overline{f_i(m + pk)}v(k).$$

These convolution operators will be called *filters* by analogy with quadrature mirror filters in the case $p = 2$. They have the following properties:

LEMMA 2.1. For $i, j = 0, 1, \ldots, p - 1$,
  (i) $F_i F_j^* = 0$,   if $i \neq j$,
  (ii) $F_i F_i^* = I$,

(iii) $F_i^* F_i$ is an orthogonal projection of l^2, and for $i \neq j$ the ranges of $F_i^* F_i$ and $F_j^* F_j$ are orthogonal, and

(iv) $F_0^* F_0 + \cdots + F_{p-1}^* F_{p-1} = I$.

PROOF. Properties (i) and (ii) follow by interchanging the order of summation:

$$
\begin{aligned}
F_i F_j^* v(k') &= \sum_m \sum_k f_i(m+pk')\overline{f_j(m+pk)}v(k) \\
&= \sum_k \left(\sum_{m'} f_i(m')\overline{f_j(m'+p[k-k'])} \right) v(k) \\
&= \sum_k \delta_{i-j}\delta_{k-k'}v(k) = \left\{ \begin{array}{ll} v(k'), & \text{if } i = j, \\ 0, & \text{if } i \neq j. \end{array} \right.
\end{aligned}
$$

For property (iii) we use (i) and (ii): $F_i^* F_i F_i^* F_i = F_i^* F_i$, and $F_i^* F_i F_j^* F_j = 0$. Orthogonality is shown by transposition: $\langle F_i^* F_i x, F_j^* F_j y \rangle = \langle F_i x, F_i F_j^* F_j y \rangle = \langle F_i x, 0 \rangle = 0$.

To prove (iv), let $m_j(\xi) = \sum_k f_j(k)e^{ik\xi}$ be the (bounded, Hölder continuous, periodic) function determined by the filter f_j, for $j = 0, \ldots, p-1$. Then $f_j(k) = \hat{m}_j(k)$ is a real number, and each $F_i^* F_i$ is unitarily equivalent to multiplication by $|m_i|^2$ on $L^2(-\pi, \pi)$.

Now Plancherel's theorem gives

$$
\int_0^{2\pi} e^{ilp\xi} m_j(\xi)\overline{m_{j'}(\xi)}\, d\xi = \sum_k f_j(k)\overline{f_{j'}(k+lp)} = \delta_{j-j'}\delta_l.
$$

In particular, $|m_j|^2$ has integral 1, and the Fourier coefficient $(|m_j|^2)\widehat{}\,(lp)$ vanishes if $l \neq 0$. This is equivalent to the average of $|m_j(\xi)|^2$ over $\{\xi, \xi + 2\pi/p, \ldots, \xi + 2\pi(p-1)/p\}$ being identically 1.

The same vanishing is true of the Fourier coefficients of the cross terms $m_j \overline{m_{j'}}$, and for those it also holds when $l = 0$. Thus, the average of $m_j(\xi)\overline{m_{j'}(\xi)}$ over $\{\xi, \xi + 2\pi/p, \ldots, \xi + 2\pi(p-1)/p\}$ vanishes identically. Hence, the conditions on the filters f_i are equivalent to the unitarity of the following matrix:

$$
\begin{pmatrix}
m_0(\xi) & m_0(\xi + \frac{2\pi}{p}) & \cdots & m_0(\xi + \frac{2\pi(p-1)}{p}) \\
\cdots & \cdots & \cdots & \cdots \\
m_{p-1}(\xi) & \cdots & \cdots & m_{p-1}(\xi + \frac{2\pi(p-1)}{p})
\end{pmatrix}
$$

But then $\sum_{k=0}^{p-1} |m_k(\xi)|^2 = 1$ for all ξ. Thus $F_0^* F_0 + \cdots + F_{p-1}^* F_{p-1}$ is unitarily equivalent to multiplication by 1 in $L^2(-\pi, \pi)$, proving (iv). $\square$

With this lemma we can decompose l^2 into mutually orthogonal subspaces $W_0^1 \perp \cdots \perp W_{p-1}^1$, where $W_i^1 = F_i^* F_i(l^2)$ for $i = 0, \ldots, p-1$. The map F_i finds the coordinates of a vector with respect to an orthonormal basis of W_i^1. One level of this decomposition is displayed in the figure below:

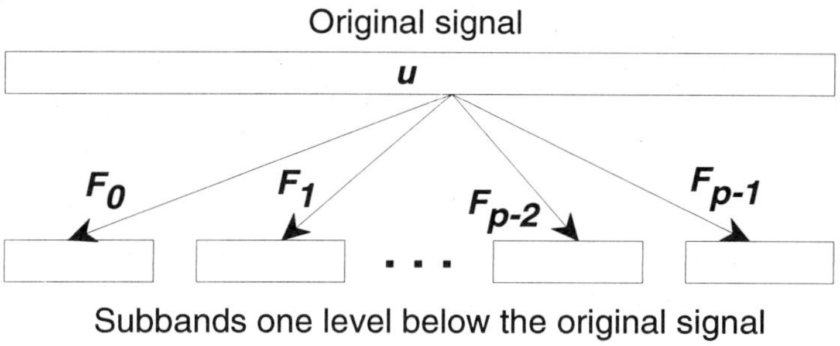

Original signal

FIGURE 2. One level of decomposition into p subspaces.

Since each $F_i W_i^1 = F_i(l^2)$ is another copy of l^2, there is nothing to prevent us from reapplying the filter convolutions recursively. At the mth stage, we obtain $l^2 = W_0^m \perp \cdots \perp W_{p^m-1}^m$, where $W_n^m = F_{n_1}^* \ldots F_{n_m}^* F_{n_m} \ldots F_{n_1}(l^2)$ and $n_m \ldots n_1$ is the radix-p representation of n. The map $F_{n_m} \ldots F_{n_1}$ transforms into standard coordinates in W_n^m. For convenience, we will introduce the notations $\mathbf{F}_n^m = F_{n_m} \ldots F_{n_1}$, and $\mathbf{F}_n^{m*} = F_{n_1}^* \ldots F_{n_m}^*$.

The subspaces W_n^m form a p-ary tree. Every node W_n^m is a parent with p daughters $W_{pn}^{m+1}, \ldots, W_{pn+p-1}^{m+1}$. The root of the tree is the original space l^2, which we may label W_0^0 for consistency. Call the whole tree $\mathbf{W}$. This tree is a partially ordered set with minimal element W_0^0, which we call the *root*. We will say that $W' \in \mathbf{W}$ is greater than $W \in \mathbf{W}$ if the unique path between W' and W_0^0 contains W. The set $\{W' : W' \geq W\}$ will be called the *descendants* of W.

Now fix m and suppose w belongs to W_n^m, where $0 \leq n \leq p^m - 1$, and $\mathbf{F}_n^m w = e_k$ is the elementary sequence with 1 in the kth position and 0's elsewhere. The collection of all such w forms an orthonormal basis of l^2 with some remarkable properties. In particular, if $p = 2$ and the filters F_0 and F_1 are taken as low-pass and high-pass quadrature mirror filters, respectively, then the spaces $W_0^m, \ldots, W_{2^m-1}^m$ are all the subbands at level m. These have been used for a long time in digital signal processing and compression. An earlier paper [**12**] described experiments with an algorithm for choosing m so as to reduce the bit rate of digitized acoustic signal transmission. This produced good signal quality at rather low bit rates.

The tree contains other orthogonal bases of W_0^0. In fact, it forms a library of bases which may be adapted to classes of functions. The tree structure allows the library to be searched efficiently for the extremum of certain cost functionals.

To every node in $\mathbf{W}$ we associate the subtree of all its descendants. Define a *graph* to be any finite subset of the nodes of $\mathbf{W}$ with the property that the union of the associated subtrees is disjoint and contains a complete level $W_0^m, \ldots, W_{p^m-1}^m$ for some m. For example, the singleton $\{W_0^0\}$ is a graph with $m = 0$. The following may be called the graph theorem.

THEOREM 2.2. *Every graph corresponds to a decomposition of l^2 into a finite direct sum of orthogonal subspaces.*

PROOF. Every graph is a finite set, of cardinality no more than p^m for the m in the definition. Fix a graph, and suppose that $W_{n_1}^{m_1}$ and $W_{n_2}^{m_2}$ are subspaces corresponding to two nodes. Without loss, suppose that $m_1 \leq m_2$. Then $W_{n_2}^{m_2}$ is contained in a subspace $W_n^{m_1}$ for some $n \neq n_1$. Since the subspaces at a given level are orthogonal, we conclude that $W_{n_2}^{m_2} \perp W_{n_1}^{m_1}$.

To show that the decomposition is complete, observe that a node contains the sum of its immediate descendent nodes, or *children*. By induction, it contains the sum of all of the nodes in its subtree. Hence a graph contains the sum of all the subspaces at some level m. But this sum is all of l^2. $\square$

COROLLARY 2.3. *Graphs are in one-to-one correspondence with finite disjoint covers of $[0, 1)$ by p-adic intervals $I_n^m = p^{-m}[n, n+1)$, $n = 0, 1, \ldots, p^m - 1$.*

PROOF. The correspondence is evidently $W_n^m \leftrightarrow I_n^m$. The subtree associated to W_n^m corresponds to all p-adic subintervals of I_n^m. The details are left to the reader. $\square$

This correspondence induces a partial order on graphs. We will say that graph u is greater than or equal to graph v if the cover associated to u is a refinement of the cover associated to v. This partial order has a minimal element $\{W_0^0\}$. For each *maximum level* $L \geq 0$ it also has a maximal element $\{W_0^L, \ldots, W_{p^L-1}^L\}$. Some example graphs are depicted in the figures below:

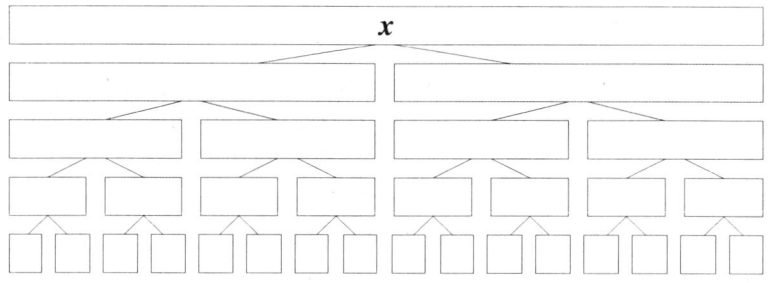

FIGURE 3. The "wavelet" graph basis, $p = 2$.

3. Analytic properties of graphs: continuous wavelet packets.

Each filter F_j (and its adjoint F_j^*) maps the class of rapidly decreasing sequences to itself. Likewise, the projections $\mathbf{F}_n^{m*}\mathbf{F}_n^m$ preserve that class. In practice, we shall consider only finite sequences in l^2. For actual computations the filters must be finitely supported as well. Convolution with such filters preserves the property of finite support. Let the support width of the filters be r, and let

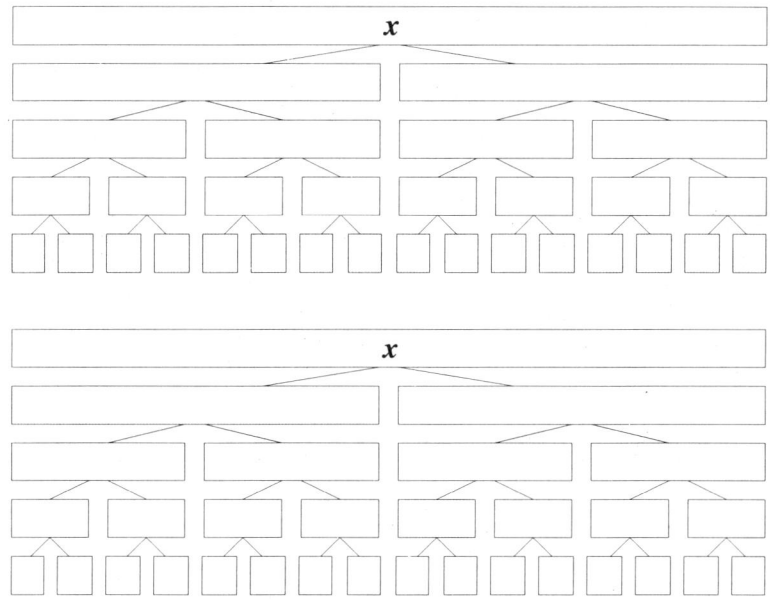

FIGURE 4. Some other examples of graph bases, $p = 2$.

z_m be the maximum width of any vector of the form $F^*_{j_1} \ldots F^*_{j_m}(e_k)$. Then $z_0 = 1$ and $z_{m+1} = pz_m + r - p$. By induction, we see that $z_m = p^m + (p^m - 1)(r - p)$.

In [**1**] we observed that the basis elements $\mathbf{F}^{m*}_n e_k$ form wave packets over **R**. Because they are superpositions of Daubechies' compactly-supported *wavelets*, we will call these basis elements *wavelet packets*. A slightly generalized paraphrase of the construction follows. Many of the basic facts we use were proved by Daubechies in [**2**].

Let w be a function defined by $\hat{w}(\xi) = \prod_{j=1}^{\infty} m_0(\xi/p^j)$, where m_0 is the analytic function defined by F_0, as above. Then w has mass 1, decreases rapidly, and is Hölder continuous, as proved in [**2**]. If m_0 is a trigonometric polynomial of degree r, then w is supported in the interval $[-r, r]$. Arranging that w has r continuous derivatives requires m_0 with degree at most $O(r)$. See [**2**] for a discussion of the constant in this relation for $p = 2$. Put $w_0^0 = w$, and define the family of wavelet packets recursively by the formula $w_{pn+j}^{m+1}(t) = \sum_{-\infty}^{\infty} f_j(i)w_n^m(pt - i)$. This produces one function w_n^m for each pair (m, n), where $m = 0, 1, \ldots$ and $n = 0, 1, \ldots, p^m - 1$.

We can renormalize the wavelet packets to a fixed scale p^L. Write

$$w_{n,m,k}^L(t) = p^{(L-m)/2} w_n^m(p^{L-m}t - k).$$

Then $w_{0,0,k}^L$ is a collection of orthonormal functions of mass $p^{L/2}$, concentrated in intervals of size $O(p^{-L})$. This makes them suitable for sampling continuous functions. Let $x(t)$ be any continuous function, and put

$$s_0^0(k) = \langle x, p^{L/2} w_{0,0,k}^L \rangle = \int_{-\infty}^{\infty} x(t) p^L w_0^0(p^L t - k)\, dt.$$

We may use $s_0^0(k)$ as a representative value of $x(t)$ in the interval $I_k^L = p^{-L}[k, k+1)$. The closeness of the approximation to values of x depends, of course, on the smoothness of x. Suppose that x is Hölder continuous with exponent ϵ. Then if t_0 is any point in I_k^L, we have

$$|x(t_0) - s_0^0(k)| = |\int_{I_k^L} (x(t_0) - x(t))\, p^L w_0^0(p^L t - k)\, dt| = O(p^{-\epsilon L}).$$

We can also take advantage of differentiability of x if we construct w_0^0 with vanishing moments. Given d vanishing moments and d derivatives of x, the approximation improves to $|x(t_0) - s_0^0(k)| = O(p^{-dL})$.

The map $x \mapsto s_0^0$ sends $L^2(\mathbf{R})$ to l^2, and pulls back the orthonormal bases of l^2 constructed in the last section. To see this, define $s_n^m(k) = \langle x, w_{n,m,k}^L \rangle$. By interchanging the order of recurrence relation and inner product, we obtain the formula $s_n^m = \mathbf{F}_n^m s_0^0$. Thus, the coordinates $s_n^m(k)$ are coefficients with respect to an orthonormal basis of W_n^m.

The resulting subspaces of $L^2(\mathbf{R})$ form a finer type of multiresolution decomposition than that of Mallat [8]. The coordinates $s_n^m(k)$ are rapidly computable. As we shall see, they contain a mixture of location and frequency information about x.

4. Ordering the basis elements.

The parameters n, m, k, L in $w_{n,m,k}^L$ have a natural interpretation as frequency, scale, position, and resolution, respectively. However, n is not monotonic with frequency, because our construction yields wavelet packets in the so-called Paley (natural, or p-adic) ordering. The following results show how to permute $n \mapsto n'$ into a frequency-based ordering.

THEOREM 4.1. We can choose rapidly decreasing filters $F_0, \ldots, F_p - 1$ such that $w_{n,m,k}^L$ is concentrated near the interval I_k^{L-m}, and $\hat{w}_{n,m,k}^L$ is concentrated near the interval $I_{n'}^m$, where $n \mapsto n'$ is a permutation of the integers.

PROOF. For the first part, we note that for any family of rapidly decreasing filters, w_0^0 decreases rapidly away from $[0, 1)$. The dilate and translate $w_{0,m,k}^L$ of this function to the interval I_k^{L-m} similarly has rapid decrease. Likewise, $w_{n,m,k}^L$ has the same concentration as $w_{0,m,k}^L$, since all the filters F_i are rapidly decreasing.

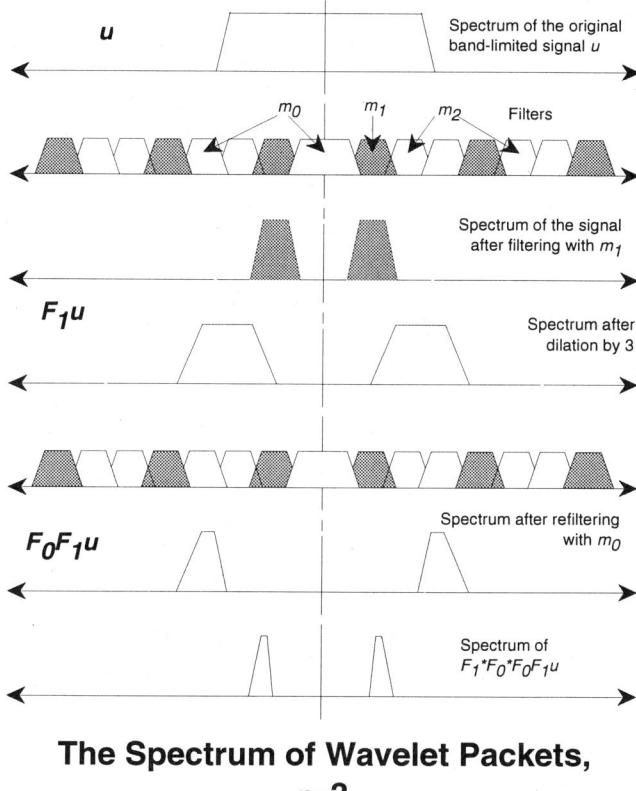

The Spectrum of Wavelet Packets,
$p=3$

FIGURE 5. Successive application of $p = 3$ convolution-decimations.

The second part may be deduced from the Fourier transform of the recurrence relation:

$$\hat{w}_{pn+j}^{m+1}(\xi) = \left(p^{-1} \sum_k f_j(k) e^{-ix\xi/p} \right) \hat{w}_n^m(\xi/p) = p^{-1} m_j(\xi/p) \hat{w}_n^m(\xi/p),$$

where m_j is the multiplier defined above. Recall that $\sum_{j=0}^{p-1} |m_j(\xi)|^2 \equiv 1$ and that $m_0(0) = 1$. Thus, the periodic functions $|m_j|^2$ form a partition of unity into p functions, with 0 being in the support of m_0 alone.

Now suppose for simplicity that we have chosen filters in such a way that $|m_j(\xi)| = \sum_{k=-\infty}^{\infty} \chi_{\pm\frac{\pi}{p}[j,j+1)}(\xi - 2\pi k)$. Such m_j may be approximated in $L^2(-\pi, \pi)$ as closely as we like by multipliers arising from exponentially decreasing filters. In this simple case, it is immediate that $\hat{w}_0^0(\xi) = m_0(\xi/p)|_{(-\pi,\pi)}$ is the characteristic function of $(-\pi, \pi)$, so that $\hat{w}_{0,0,0}^L$ is the characteristic function of $(-\pi p^L, \pi p^L)$. Likewise, $\hat{w}_{j,1,0}^L$ is the characteristic function of $\pi p^{L-1}(-j-1, -j] \cup \pi p^{L-1}[j, j+1)$. From the recurrence relation, we see that $\hat{w}_{n,m,0}^L$ will be the characteristic function of the union of the intervals $\pm\pi p^{L-m}[n', n'+1)$, where $n \mapsto n'$ is a permutation. These intervals cover $p^L(-\pi, \pi)$ as $n = 0, \ldots, p^m - 1$. This arrangement of frequencies is depicted in Figure 5. The permutation $n \mapsto n'$ is given by the recurrence relation

$$n' = n, \quad \text{if } n = 0, \ldots, p-1; \quad (np+j)' = \begin{cases} n'p + j, & \text{if } n' \text{ is even,} \\ n'p + (p-1) - j, & \text{if } n' \text{ is odd.} \end{cases}$$

Write n_j for the jth digit of n in radix p, numbering from the least significant. Set $n_m = 0$ if n has fewer than m digits. Then the recurrence relation implies that $n_j = \pi(n'_{j+1}, n'_j)$, where

$$\pi(x,y) = \begin{cases} y, & \text{if } x \text{ is even,} \\ p-1-y, & \text{if } x \text{ is odd.} \end{cases}$$

For each value of the first variable, π is a permutation of the set $\{0, \ldots, p-1\}$ in the second variable. Thus the map $n' \mapsto n$ and its inverse $n \mapsto n'$ are permutations of the integers. It is not hard to see that these are permutations of order 2 if p happens to be odd. Otherwise they have infinite order, as may be seen by considering an increasing sequence of integers n' all of which have only odd digits in radix p. $\square$

COROLLARY 4.2. With filters $F_0, \ldots, F_p - 1$ chosen as above, we can modify the recurrence relation for $w_{n,m,k}^L$ such that $\hat{w}_{n,m,k}^L$ is concentrated near the interval I_n^m.

PROOF. Simply reorder the functions w_n^m by using the alternative recurrence relation:

$$w_{pn+j}^{m+1}(t) = \begin{cases} \sum_k f_j(k) w_n^m(pt - k), & \text{if } n \text{ is even,} \\ \sum_k f_{p-1-j}(k) w_n^m(pt - k), & \text{if } n \text{ is odd.} \end{cases}$$

Since we are enforcing $n = n'$ at each level m, we are composing with the permutation defined above. Of course, this algorithm has complexity identical to the original. $\square$

5. Periodic filters and bases for $\mathbf{R}^d$.

A sampled periodic function may be represented as a vector in $\mathbf{R}^d$ for some d. In this case let p be any factor of d. Introduce as filters a family of p vectors $\{\tilde{f}_i \in \mathbf{R}^d : i = 0, \ldots, p-1\}$. These are obviously summable. Suppose in addition that they are orthogonal as periodic discrete functions, i.e., that $\sum_{m=1}^d \tilde{f}_i(m)\tilde{f}_j(m + kp \mod d) = \delta_{i-j}\delta_k$.

Let the associated convolution operators be $\{\tilde{F}_0, \ldots, \tilde{F}_{p-1}\}$, defined as above by

$$\tilde{F}_i : \mathbf{R}^d \to \mathbf{R}^{d/p}, \ \tilde{F}_i v(k) = \sum_{m=1}^{d} \tilde{f}_i(m + pk \mod d) v(m), \quad \text{for } k = 1, 2, \ldots, d/p,$$

$$\tilde{F}_i^* : \mathbf{R}^{d/p} \to \mathbf{R}^d, \ \tilde{F}_i^* v(m) = \sum_{k=1}^{d/p} \overline{\tilde{f}_i(m + pk \mod d)} v(k), \quad \text{for } m = 1, 2, \ldots, d.$$

The reduction modulo d is intentionally emphasized. These operators satisfy conditions similar to those of aperiodic filters:

LEMMA 5.1.
   (i) $\tilde{F}_i \tilde{F}_j^* = 0,$ if $i \neq j$,
   (ii) $\tilde{F}_i \tilde{F}_i^* = I_{d/p}$
   (iii) $\tilde{F}_i^* \tilde{F}_i$ is a rank d/p orthogonal projection on $\mathbf{R}^d$, and for $i \neq j$ the ranges of $\tilde{F}_i^* \tilde{F}_i$ and $\tilde{F}_j^* \tilde{F}_j$ are orthogonal,
   (iv) $\tilde{F}_0^* \tilde{F}_0 + \cdots + \tilde{F}_{p-1}^* \tilde{F}_{p-1} = I_d$
where I_d is the identity on $\mathbf{R}^d$.

PROOF. The proof is nearly identical with the one in the aperiodic case. □

The decomposition suggested by equation (iv) may be recursively applied to the p subspaces $\mathbf{R}^{d/p}$ to generate *periodic wavelet packets*. We must extend the action of the filter family to $\mathbf{R}^{d/p}$ in the natural way. For $d = p_1 \ldots p_L$ and $0 \leq n < d$, we have a unique representation $n = n_1 + n_2 p_1 + n_3 p_2 p_1 + \cdots + n_L p_{L-1} \ldots p_1$, where $0 \leq n_i < p_i$. This defines a one-to-one correspondence between $\{0, \ldots, d-1\}$ and an index set of L-tuples $I = \{ (n_1, \ldots, n_L) : 0 \leq n_i < p_i \}$. We can construct a basis of $\mathbf{R}^d$ whose elements are indexed by I. For $n = (n_1, \ldots, n_L) \in I$, define $\tilde{\mathbf{F}}_n^L = \tilde{F}_{n_L}^L \ldots \tilde{F}_{n_1}^1$, where $\tilde{F}^i$ is a family of p_i periodic filters. Then $\tilde{\mathbf{F}}_n^{L*} \tilde{\mathbf{F}}_n^L$ is an orthogonal projection onto a 1-dimensional subspace of $\mathbf{R}^d$. This is shown by induction on the rank in (iii). Now let W_n^L be the range of this projection. The collection $\{u_n = \tilde{\mathbf{F}}_n^{L*} 1 : n \in I\}$ of standard basis vectors of W_n^L will be an orthonormal basis of $\mathbf{R}^d$, and the map $\tilde{\mathbf{F}}_n^L : \mathbf{R}^d \to \mathbf{R}$ gives the component in the u_n direction. Some examples of periodic wavelet packets are depicted in Figures 6 and 7.

   These periodic wavelet packets are more useful in practice than the aperiodic wavelet packets we first considered, because the number of coefficients produced by the periodic wavelet packet transform is no more than the original number of signal samples. Of course, this advantage is balanced by the requirement that we treat the original signal as periodic. Notice that the periodized wavelet packets with consecutive indices $2n-1$ and $2n$ differ only by a shift. This shift, which is always close to 1/4 period, bears a simple relationship to the binary expansion of n, but its main significance is that there are really only $N/2$ distinct frequencies in a collection of N periodized wavelet packets. This is a reflection of Nyquist's

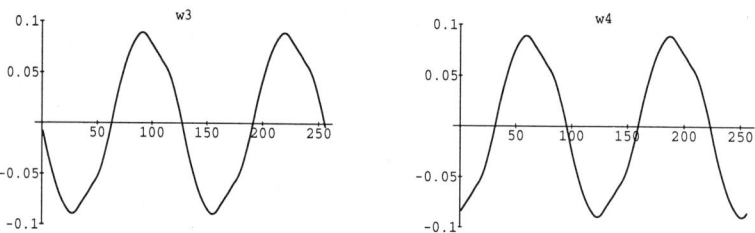

FIGURE 6. Periodized wavelet packets, $p = 2$, sequencies 3 and 4.

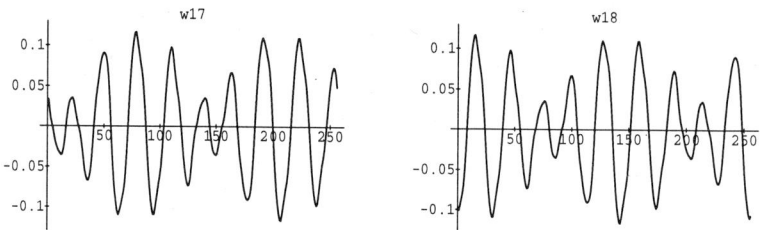

FIGURE 7. Periodized wavelet packets, $p = 2$, sequencies 17 and 18.

theorem, which states that if we sample a periodic function N times within a period, then we can only distinguish frequencies up to $N/2$.

As before, we are not limited to the basis defined by the index set I. Products of fewer than L filters form orthogonal projections onto a tree of subspaces of $\mathbf{R}^d$. A node arising from a product of m filters will correspond to the subspace $W_n^m = \tilde{\mathbf{F}}_n^{m*}\tilde{\mathbf{F}}_n^m\mathbf{R}^d$, where $n = n_1 + \cdots + n_m p_{m-1} \ldots p_1$ indexes a composition of m filters. The tree will be nonhomogeneous in general, although all nodes i levels from the root will have the same number p_i of children. Define a *nonhomogeneous graph* as a finite union of nodes whose associated subtrees form a disjoint cover of some level $m \leq L$. Then the following graph theorem and its corollary hold for this tree of subspaces:

THEOREM 5.2. *Every nonhomogeneous graph corresponds to an orthogonal decomposition of* $\mathbf{R}^d$.

COROLLARY 5.3. *Graphs are in one-to-one correspondence with finite disjoint covers of* $[0, 1)$ *by intervals of the form* $I_n^m = (p_1 \ldots p_m)^{-1}[n, n+1)$.

This correspondence also induces a partial order on nonhomogeneous graphs. We say that u is greater than or equal to v if the cover associated to u is a refinement of the cover associated to v. This partial order has both a minimal element $\{W_0^0\}$ and a maximal element $\{W_0^L, \ldots, W_{p_1 \ldots p_L - 1}^L\}$. Any permutation of the factors of d gives a (possibly different) set of bases.

6. Concentration criteria for the best-adapted basis.

Define an *additive information cost function* on l^2 to be any functional $M : l^2 \to \mathbf{R}^+$ satisfying

$$\text{(i)} \quad M(\{x_i\}) = \sum_i m(|x_i|) \quad \text{for some function } m = m(t),$$

$$\text{(ii)} \quad M(\{0\}) = 0.$$

Some useful information cost functions are the threshold counting norm $\#\{k : |x_k| > \epsilon\}$ and the bit-length norm $\sum_k \log(1 + |x_k|/\epsilon)$.

Let U be a finite library of orthonormal bases of l^2. A vector x has coefficients x_u in the basis $u \in U$, where $x_u(k) = \langle x, u(k) \rangle$ for $u(k) \in u$. The information cost of a particular representation may be measured by $M(x_u)$. Also, the information contained in the the choice of u is $\log |U|$, where $|U|$ is the cardinality of the library. Define a *best-basis* for x as any element $u \in U$ for which $M(x_u)$ is minimal. The *best-basis information cost* of x in the library U is therefore $M(x_u) + \log |U|$.

Our goal is to find an efficient algorithm for reducing the information cost of vectors in a class. The library U, which depends on the class, should be very large but easy to search. It takes a naïve algorithm $O(|U|)$ operations to find the least-information representation of a fixed vector x. This procedure is inefficient because it requires a global reëvaluation of the information cost for each basis in the library.

We can evidently reduce the information $M(x \oplus y)$ by reducing $M(x)$ and $M(y)$ individually. Such a procedure is local in the following sense. Suppose that 2 orthonormal bases u and v in a library U partially coincide, and we write $x_u = x_{u \cap v} \times x_{u'}$ and $x_v = x_{u \cap v} \times x_{v'}$. Then $M(x_u) < M(x_v) \iff M(x_{u'}) < M(x_{v'})$. The decomposition and individual reduction of M may then be reapplied to the pieces $x_{u'}$ and $x_{v'}$.

Let $\mathbf{W}$ be a tree of subspaces of l^2. If $W \in \mathbf{W}$ is a node, then the associated coordinate map is $\mathbf{F}_W$, and the associated orthogonal projection can be denoted by $\mathbf{F}_W^* \mathbf{F}_W$. Every node W may be regarded as the Cartesian product of its daughters, or more generally of the elements of any graph of its descendants. Let U be the library of bases corresponding to graphs in $\mathbf{W}$. Then each element of U has a unique factorization into a Cartesian product of the standard bases of the subspaces in the associated graph. Namely, if x is a sequence and if u is a basis corresponding to some graph $G \subset \mathbf{W}$, then $x_u = \times \prod_{W \in G} \mathbf{F}_W x$. Consequently, $M(x_u) = \sum_{W \in G} M(\mathbf{F}_W x)$. To find the best basis, we must choose a graph of those subspaces W which contribute the least information. But this large choice may be factored into a sequence of small subchoices.

Recall now our definition of a partial ordering on graphs through the tree $\mathbf{W}$, which is inherited from our definition of the partial order on trees. We must keep track of the lowest achievable measure of information M as we progress down the tree to its root. This may be defined inductively. We must suppose now that the tree is finite with L levels, so that its set of graphs has a maximal element. Let G_L be this maximal graph in $\mathbf{W}$, and for $W \in G_L$ set $M_W^*(x) = M(\mathbf{F}_W x)$. Then for $W \in \mathbf{W}$ let $M_W^*(x) = \min\{M_V^*(x) : V \geq W\}$. This functional M_W^* records the minimum value of $M(\mathbf{F}_W x_u)$ achievable by bases of W coming from graphs through subtrees above node W. Evidently, if W_0 is the root (or minimal) node of $\mathbf{W}$, then $M_{W_0}^*(x) = M(x_{u_{\min}})$ is the best-basis measure of information for the vector x.

The search algorithm may now be described. Mark all maximal nodes $W \in G_L$ as "kept." Suppose now that node W has children $W_1, \ldots, W_n$. Mark W as "kept" if $M_W^*(x) \leq M_{W_1}^*(x) + \cdots + M_{W_n}^*(x)$; otherwise mark it as "not kept". Namely, keep W if including it reduces M^*. Observe that we can compute $M_W^*(x) = \min\{M(F_W x), M_{W_1}^*(x) + \cdots + M_{W_n}^*(x)\}$ without having to search the entire subtree above W. We may proceed down the tree to the root, at which point all the nodes in the tree have been marked either as "kept" or as "not kept." We claim the following:

PROPOSITION 6.1. The union of the minimal "kept" nodes is a graph corresponding to the best-basis representation of x.

PROOF. That it is a graph is clear by induction. Every minimal "kept" node W is the root of a subtree containing some of the maximal nodes of $\mathbf{W}$. This set of subtrees is disjoint, since if two subtrees intersect then one must contain the other and so their roots cannot both be minimal. The union of these disjoint subtrees covers all of the maximal nodes of $\mathbf{W}$, which form a complete level of the tree.

So call this graph G. Notice that the sum $\sum_{W \in G} M_W^*(x)$ over the minimal "kept" nodes $W \in G$ is equal to $M_{W_0}^*(x)$, where W_0 is the root of $\mathbf{W}$. By the remarks immediately above the proposition, this is the minimum achievable information cost. $\square$

7. Operations required to find the best basis.

We may count the operations in our search algorithm above as follows. Let $E(W)$ be the number of operations required to evaluate $M(F_W x)$, and let $D(W)$ be the number of children of the node $W \in \mathbf{W}$. Then it will require $\sum_{W \in G_L} E(W) + \sum_{W \notin G_L} [E(W) + D(W)]$ operations to construct the functional M^* on the tree and to mark the appropriate nodes as "kept." Finding the minimal "kept" nodes requires a depth-first search of $\mathbf{W}$, which takes at most $|\mathbf{W}|$ operations.

For definiteness, consider the example of a homogeneous p-adic tree generated by periodizing a family of p filters. Suppose we start with a vector x of $N = p^L$ components and develop the tree of its representations $\mathbf{W}$ as far as we can, namely L levels. We can label the nodes W_n^m as before. We observe that $\mathbf{F}_n^m x$ has p^{L-m} components so that $E(W_n^m) = cp^{L-m}$, where c is the number of operations required per non-zero coefficient to compute M. This tree has p^m nodes at level m. $D(W) = p$ for all $W \in \mathbf{W}$, so it requires $\sum_{n=0}^{p^L-1} cp^0 + \sum_{m=0}^{L-1} \sum_{n=0}^{p^m-1} [cp^{L-m}+p] = cp^L + p(p^L-1)/(p-1) = O(p^L) = O(N)$ operations to build M^*. Then $|\mathbf{W}| = \sum_{m=0}^{L} p^m = (p^{L+1}-1)/(p-1) = O(p^L) = O(N)$, so that the entire search takes $O(N)$ operations for a periodic vector of length N.

The library of graphs through a tree grows rapidly with the number of levels in the tree. Let $|U_L|$ be the number of bases in a p-adic tree of L levels. Then $|U_L|$ satisfies the recurrence $|U_{L+1}| = 1 + |U_L|^p$, which is easily estimated as greater than $2^{p^{L-1}} = 2^{N/p}$. We list the first 7 values $|U_0|, \dots, |U_6|$ for $p = 3$:

```
1, 2, 9, 730, 389017001, 58871587162270593034051002,
204040901322752673844230437877671861543858084850895762746141813554591014612009
```

By contrast, one may also list the number of operations required to find the best basis representation and information of a vector of length $N = 3^L$ in a tree of $L = 0, \dots, 6$ levels. We shall suppose that evaluating M requires 3 operations per coefficient, so that the operation count is at most $\frac{3}{2}(3^{L+1} - 1)$.

$$\{3, 12, 39, 120, 363, 1092, 3279\}$$

This is a good example of combinatorial explosion tamed by an efficient search algorithm.

8. The entropy criterion and estimates.

Let $x \in l^2$ and denote by $\|x\|$ the usual norm: $\|x\|^2 = \sum_k |x_k|^2$. Then the sequence defined by $|x_k|^2/\|x\|^2$ gives a probability distribution of the energy of x. This distribution has a Shannon entropy which we shall denote by $\mathcal{H}(x) = -\sum_k (|x_k|^2/\|x\|^2) \log(|x_k|^2/\|x\|^2)$, where the summand is interpreted as 0 for any $x_k = 0$. This entropy is a well-known measure of the information of a distribution.

We may also use the $L^2 \log L^2$ norm rather than entropy. Denote this by $H(x) = -\sum_k |x_k|^2 \log |x_k|^2$ with the same convention for the case $x_k = 0$. Note that

$$\mathcal{H}(x) = H(x)\|x\|^{-2} + \log \|x\|^2$$

H is an additive measures of information. $\mathcal{H}$ is not, but the above relation guarantees that whenever $\|x\| = \|y\|$, we have $\mathcal{H}(x) < \mathcal{H}(y) \iff H(x) < H(y)$. Suppose that u is a best basis for x with respect to H. It is clear that this will also be the element for which the more classical $\mathcal{H}(x_u)$ is minimal, and also that for which $\exp \mathcal{H}(x_u)$ is minimal. The exponential of entropy has the following suggestive property:

LEMMA 8.1. If $x \in l^2$ is a sequence of 0's and 1's, then $\exp \mathcal{H}(x)$ is the number of 1's in the sequence.

Notice that $\exp \mathcal{H}(x) = \|x\|^2 \exp\left(H(x)\|x\|^{-2}\right)$.

Suppose now that $x \in l^2$ is any sequence, and we project it onto a sequence y_ϵ defined for $\epsilon \geq 0$ by

$$y_k = \begin{cases} 0, & \text{if } |x_k|^2 < \epsilon \exp\left(-H(x)\|x\|^{-2}\right), \\ x_k, & \text{otherwise.} \end{cases}$$

Then there will be at most $\exp\left(-H(x)\|x\|^{-2}\right)/\epsilon$ nonzero terms in y. The energy error $\|x - y\|^2$ will be the sum of the squares of the omitted terms.

9. Existence and construction of filters.

We can construct finitely supported filters of any support length greater that p. Longer support lengths allow more degrees of freedom. Let M be a positive integer and consider the problem of finding filters of length pM, i.e., p trigonometric polynomials $m_0, \ldots, m_{p-1}$ of degree pM for which the above matrix of values of m_j is unitary. By a construction similar to Pollen's in [11], this is equivalent to finding an element of the group $SU(p, \mathbf{C}[z, 1/z])$ which is the product of M inverse factors.

Given any pair P, Q of (perfect reconstruction) quadrature mirror filters, we can build a family of $p = 2^q$ filters by taking all distinguishable compositions of P and Q of length q. Alternatively, we can take all distinguishable products of q filters. This method serves to build filters for q-dimensional signals. Given a signal $s = s(x) = s(x_1, \ldots, x_q)$, and $J = j_q \ldots j_1$ radix 2, we can define 2^q filters F_J by taking a one-dimensional filter for each dimension: $F_J s(x) = \sum_{k_1, \ldots, k_q} f_{j_1}(k_1 + px_1) \ldots f_{j_q}(k_q + px_q) s(k_1, \ldots, k_q)$. Such filters are useful for image processing and matrix multiplication.

Gopinath and Burrus [4] have given a construction of "multiplicity p" wavelets similar to the one in this paper. Their scheme for generating filter families is based on "cosine modulation," and they provided examples of filter families with Hölder regularity. In practice it is desirable to have smooth basis elements, since a certain degree of smoothness (one derivative in L^2) is needed to have finite variance in frequency. We can define a smoothness property for filter sequences:

DEFINITION 9.1. A summable sequence f is a *smooth filter* (of degree $d \leq \infty$) if there is a nonzero solution ϕ in $L^1(\mathbf{R}) \cap L^2(\mathbf{R}) \cap C^d(\mathbf{R})$ to the functional equation

$$\phi(x) = p^{1/2} \sum_m f(m)\phi(px + m).$$

Daubechies has shown in [2] that finitely supported filters of any degree of smoothness may be constructed in the case $p = 2$. An obvious consequence is that smooth filters exist in the case $p = 2^q$. For arbitrary p, Lundberg and Welland [7] give a construction of p-families of filters whose wavelet packets are m-differentiable, where p and m are arbitrary.

References

1. Ronald R. Coifman, Yves Meyer, and M. Victor Wickerhauser, *Wavelet analysis and signal processing*, Wavelets and Their Applications (Boston) (M. B. Ruskai et al., ed.), Jones and Bartlett, Boston, 1992, ISBN 0-86720-225-4, pp. 153–178.
2. Ingrid Daubechies, *Orthonormal bases of compactly supported wavelets*, Communications on Pure and Applied Mathematics **XLI** (1988), 909–996.
3. Michael Frazier, Bjørn Jawerth, and Guido Weiss, *Littlewood–Paley Theory and the Study of Function Spaces*, CBMS Regional Conference Series (Providence), American Mathematical Society, Providence, 1990.
4. R. A. Gopinath and C. S. Burrus, *Wavelet transforms and filter banks*, Wavelets–A Tutorial in Theory and Applications (Boston) (C. K. Chui, ed.), Academic Press, Boston, 1992, ISBN 0-12-174590-2, pp. 603–654.
5. A. Grossman and J. Morlet, *Decomposition of Hardy functions into square-integrable wavelets of constant shape*, SIAM J. Math. Anal. **8** (1985), 4.
6. Pierre Gilles Lemarié, *Fonctions a support compact dans les analyses multi-résolutions*, Revista Matemática Iberoamericana **7** (1991), 157–182.
7. M. Lundberg and G. Welland, *Construction of compact p-wavelets*, J. Constructive Approximation (1992), to appear.
8. Stephane G. Mallat, *A theory for multiresolution signal decomposition: The wavelet decomposition*, IEEE Transactions on Pattern Analysis and Machine Intelligence **11** (1989), 674–693.
9. Yves Meyer, *De la recherche pétrolière à la géometrie des espaces de Banach en passant par les paraproduits*, Seminaire équations aux dérivées partielles (1985–1986), preprint, École Polytechnique, Palaiseau.
10. _____, *Ondelettes et Opérateurs I,II*, Hermann, Paris, 1990.
11. David Pollen, *Parametrization of Compactly Supported Wavelets*, Aware, Inc., Cambridge, Mass., 1989, preprint AD890503.1.4.
12. M. Victor Wickerhauser, *Acoustic signal compression with wavelet packets*, Wavelets–A Tutorial in Theory and Applications (Boston) (C. K. Chui, ed.), Academic Press, Boston, 1992, ISBN 0-12-174590-2, pp. 679–700.

DEPARTMENT OF MATHEMATICS, WASHINGTON UNIVERSITY, CAMPUS BOX 1146, ONE BROOKINGS DRIVE, ST. LOUIS, MISSOURI 63130

E-mail: victor@math.wustl.edu

Proceedings of Symposia in Applied Mathematics
Volume **47**, 1993

Nonlinear Wavelet Methods for Recovery of Signals, Densities, and Spectra from Indirect and Noisy Data

DAVID L. DONOHO

ABSTRACT. We describe wavelet methods for recovery of objects from noisy and incomplete data. The common themes: (a) the new methods utilize nonlinear operations in the wavelet domain; (b) they accomplish tasks which are not possible by traditional linear/Fourier approaches to such problems. We attempt to indicate the heuristic principles, theoretical foundations, and possible application areas for these methods. Areas covered: (1) Wavelet De-Noising. (2) Wavelet Approaches to Linear Inverse Problems. (4) Wavelet Packet De-Noising. (5) Segmented Multi-Resolutions. (6) Nonlinear Multi-resolutions.

1. Introduction.

With the rapid development of computerized scientific instruments comes a wide variety of interesting problems for data analysis and signal processing. In fields ranging from Extragalactic Astronomy to Molecular Spectroscopy to Medical Imaging to Computer Vision, one must recover a signal, curve, image, spectrum, or density from incomplete, indirect, and noisy data.

What can wavelets contribute to this already intensely developed and rapidly advancing field? As it turns out quite a lot – both in theory and practice. In this paper we will give a brief discussion of several contributions.

Wavelet shrinkage. Wavelet shrinkage refers to reconstructions obtained by wavelet transformation, followed by shrinking the empirical wavelet coefficients towards zero, followed by inverse transformation.

1991 *Mathematics Subject Classification.* Primary 62A99, 62M10, 62M20, 62P99; Secondary 46B10.

Partially supported by NSF DMS 92-09130 (Stanford), and by ONR N00014-92-0066 (Statistical Sciences, Inc.).

Figure 1 gives an illustration of this method in action. An NMR signal is transformed, thresholded, and inverse transformed. The result has a noise-free visual appearance; this has been achieved without broadening features. Sections 2-4 below deal with wavelet shrinkage, its heuristic basis, and its diverse applications.

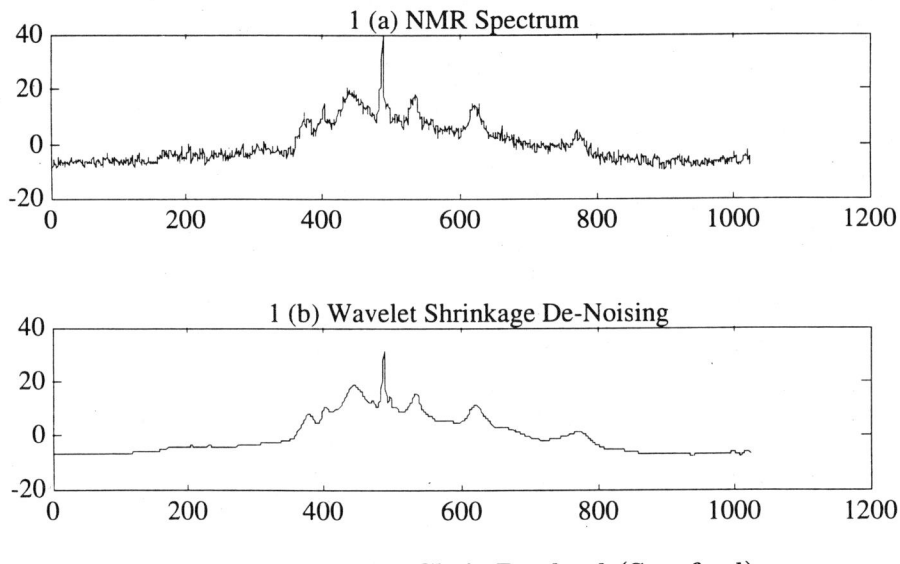

Data Provided by Chris Raphael (Stanford)

FIGURE 1.

Wavelet solution of linear inverse problems. Often scientific data are indirectly observed, as well as noisy; either the object is blurred (e.g. by a convolution operator or "point-spread function"), or else it is observed in an entirely different domain (think of the Radon transform, which gives data on line integrals of the object rather than the object itself). Such problems are typically ill-posed, in that naive attempts to undo the blurring or indirection give, in the presence of even small amounts of noise, completely useless reconstructions.

Recently the author [**7**] has proposed a Wavelet-Vaguelette Decomposition of inverse problems. Using this, one transforms the noisy, blurred data, using vaguelettes, into the wavelet domain, then thresholds the wavelet coefficients, and then applies an inverse wavelet transform. Figure 2 shows the use of this method in operation on a deconvolution problem. The improvement over naive deconvolution is evident.

Sections 5 and 6 below discuss this approach to inverse problems in more detail.

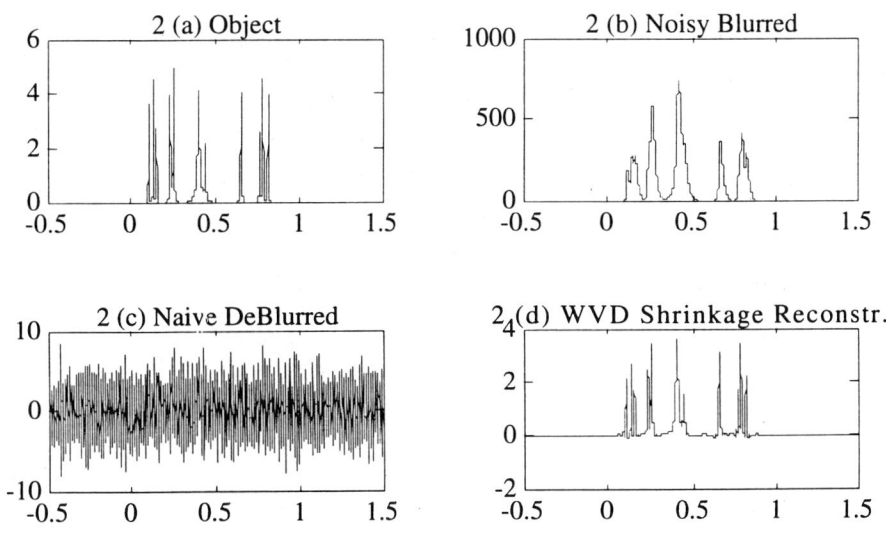

FIGURE 2.

Wavelet optimality. Wavelet methods for de-noising and de-blurring have recently attracted considerable interest; the author is aware of efforts, in fields ranging from medical imaging to synthetic aperture radar, where some variant of thresholding of wavelet coefficients is being tried. A complement to such promising empirical work is recent theoretical work which shows that from a variety of points of view these simple nonlinear wavelet methods outperform the traditional linear methods such as splines, Fourier series, and kernel-based smoothers. Some of this theoretical work, and the relation with mathematical results on wavelet bases, is described in section 7 below.

Wavelet packet de-noising. Wavelet bases are not well-suited to representing objects containing sinusoidal oscillations of moderate duration. Coifman and Meyer [3] have introduced *local cosine* bases and Coifman, Meyer, and Wickerhauser [4] *wavelet packet* bases. In constructing these bases the technical ideas underlying the wavelet transformation are deployed in different ways, leading to a large number of orthogonal basis, each one a serious alternative to classical Fourier analysis and classical wavelet basis. These transforms are better suited than wavelets for certain specific problems; an example might be in the analysis of acoustic phenomena consisting of moderate-duration damped sinusoids. There are many such transforms, though, and it is important to select those which are well-adapted to the signal at hand. Coifman and Wickerhauser [5] have introduced a method of selecting among all these transforms for the one that minimizes a certain measure of the "entropy" of the sequence, leading to a transform which in some rhetorical sense renders the data maximally simple.

Transposing these ideas into a statistical setting leads to the question of selecting a best basis for de-noising a given dataset. In section 8.1 below, we describe a method based on Stein's Unbiased Risk Estimate. Figure 3 shows the results in recovering a signal which is a superposition of moderate-duration oscillatory phenomena, from data containing both signal and white noise. Reconstruction by denoising in the adaptively-selected wavelet packet basis is much better than in the wavelet basis.

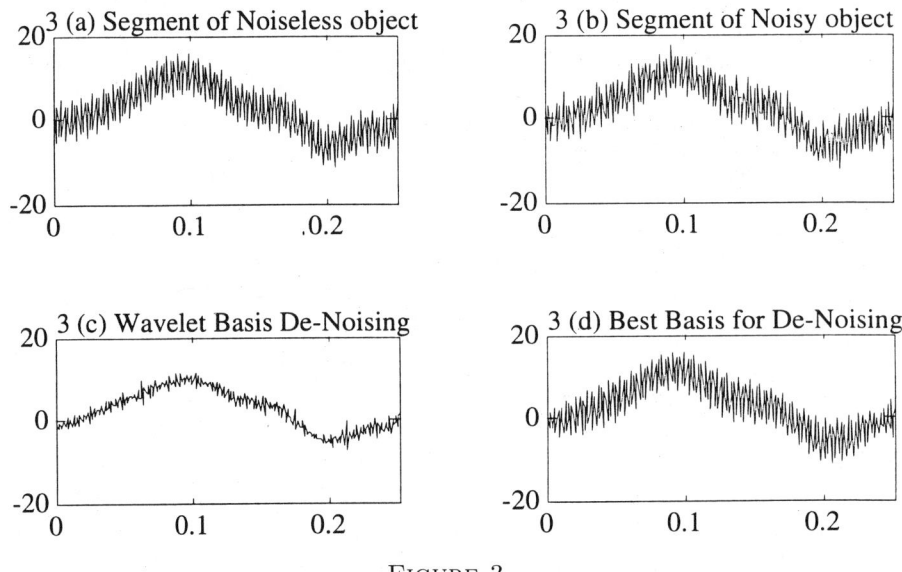

FIGURE 3.

Minimum entropy segmentation. Wavelet methods are often used in the analysis of objects containing edges – for example in 2-d image processing. Wavelets behave well but not ideally in the presence of such 2-d edges. In section 8.2 we describe a response to this, based on defining edge-preserving multiresolution operators and corresponding edge-adapted wavelet bases. The issue of adapting to edges in the 2-d transform is then, in principle, simply one of selecting that edge-adapted basis which optimally compresses the object at hand. Such a selection may be obtained by a minimum entropy criterion (noiseless data), or else by minimizing the SURE (noisy data). As a side benefit, denoising in the selected basis does not erode the edges present in the images of the object, as denoising in the wavelet basis is sometimes said to do.

Nonlinear multi-resolutions. Wavelet methods are sometimes used in the analysis of data contaminated by severe outliers. In section 8.3 we describe a response to such outliers, based on a nonlinear multiresolution analysis centered around L^1 or median fits. When the data are contaminated by extremely long-tailed error distributions, such as the Cauchy distribution, such nonlinear multiresolutions provide plausible reconstructions in cases where standard linear multiresolutions behave horribly.

Overview. In this paper we aim only to show that wavelets and associated ideas can make serious contributions to problem areas where there is already a considerable amount of interest, and to show that wavelets and associated ideas open up totally new questions in other areas. There are many other applications of wavelets in data analysis and signal processing, but we limit ourselves here to those areas where the author is directly involved. We attempt to reference a variety of work on wavelets in reconstruction and recovery, so that the reader may also find out about what others are doing. The wide variety of activities in the areas we discuss makes us hopeful that wavelets will soon have a large impact on the way in which scientists and engineers treat noisy and indirect observations.

2. De-noising by soft-thresholding.

Suppose we are interested in a function $f(t)$ on the unit interval $t \in [0, 1]$ and we have $n = 2^{J+1}$ data $y_i = f(t_i) + \sigma z_i$, $i = 1, \ldots, n$; here the t_i are equispaced and the z_i a white noise. Donoho and Johnstone [**12**] propose a three step method for recovery of $f(t)$.

(1) Perform the pre-conditioned, interval-adapted, pyramid wavelet filtering of Cohen, Daubechies, Jawerth, and Vial [**2**] to the data $\beta_{J+1,k} = y_k/\sqrt{n}$, yielding noisy wavelet coefficients $w_{j,k}$, $j = j_0, \ldots, J$, $k = 0, \ldots, 2^j - 1$.

(2) Apply the soft-threshold nonlinearity $\eta_t(w) = sgn\,(w)(|w| - t)_+$ to the noisy empirical wavelet coefficients, with threshold $t = \sqrt{2 \log(n)} \sigma / \sqrt{n}$, yielding estimates $\hat{\alpha}_{j,k}$.

(3) Set all wavelet coefficients $\hat{\alpha}_{j,k} = 0$ for $j > J$, invert the wavelet transform, producing the estimate $\hat{f}(t)$, $t \in [0, 1]$.

This method shrinks the empirical wavelet coefficients towards zero. Statisticians consider this an example of multivariate shrinkage estimates, e.g. Efron and Morris [**18**], Stein [**29**].

To see how this works, we take four functions, *Blocks*, *Bumps*, *HeaviSine*, and *Doppler*, illustrated in Figure 4. Here $n = 2048 = 2^{11}$. Noisy versions are depicted in Figure 5. Reconstructions by the method are depicted in Figure 6. The reconstructions have two properties.

(1) The noise has been almost entirely suppressed.

(2) Features sharp in the original remain sharp in reconstruction.

This behavior is very different from traditional linear methods of smoothing, which achieve noise suppression only by broadening features significantly. For comparison, Figures 7 and 8 show the results of two state-of-the-art adaptive

DAVID L. DONOHO

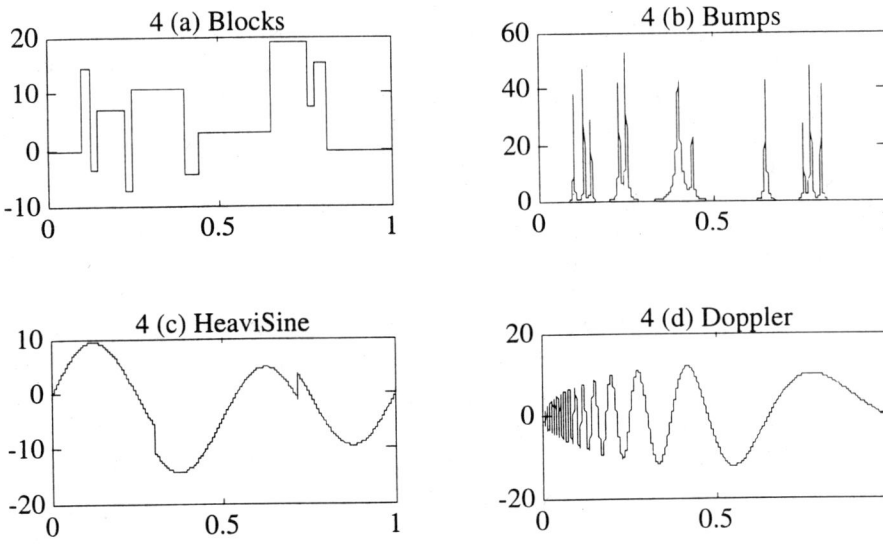

FIGURE 4.

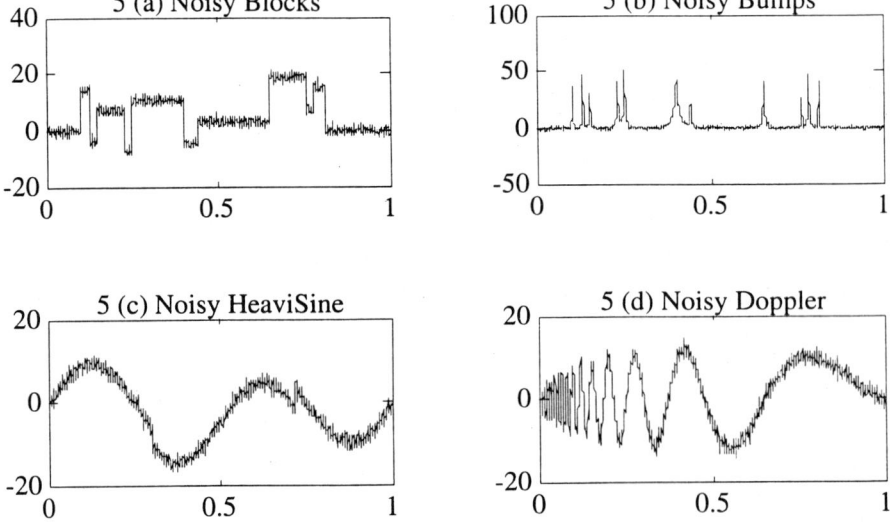

FIGURE 5.

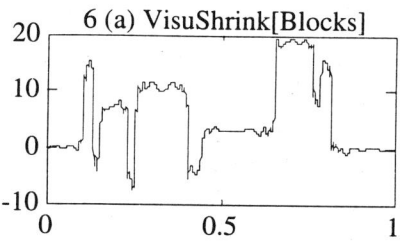

6 (a) VisuShrink[Blocks]

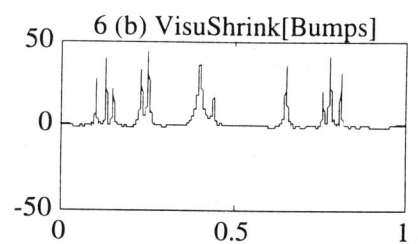

6 (b) VisuShrink[Bumps]

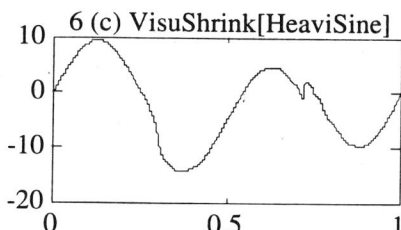

6 (c) VisuShrink[HeaviSine]

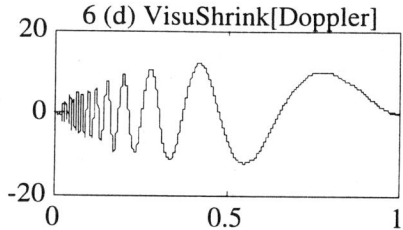

6 (d) VisuShrink[Doppler]

FIGURE 6.

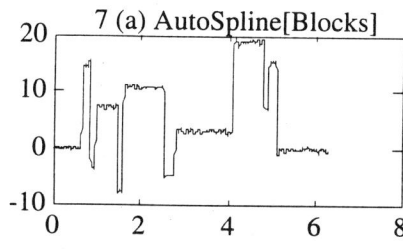

7 (a) AutoSpline[Blocks]

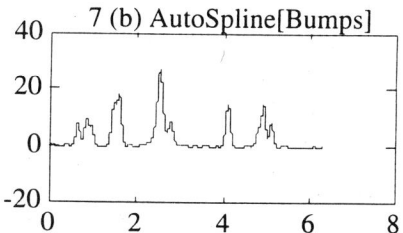

7 (b) AutoSpline[Bumps]

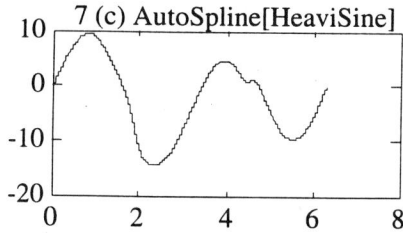

7 (c) AutoSpline[HeaviSine]

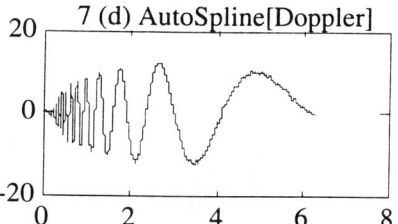

7 (d) AutoSpline[Doppler]

FIGURE 7.

linear smoothers, one based on fitting splines under tension with adaptively chosen tension parameter, and one based on truncating the empirical Fourier series with adaptively chosen truncation. (Adaptation using Stein's Unbiased Estimates of Risk [**29**]). The adaptive spline under tension suppresses noise, but at the expense of significantly broadening, and in fact erasing, certain features. The adaptive Fourier Series estimate leaves features sharp, but does not really suppress the noise.

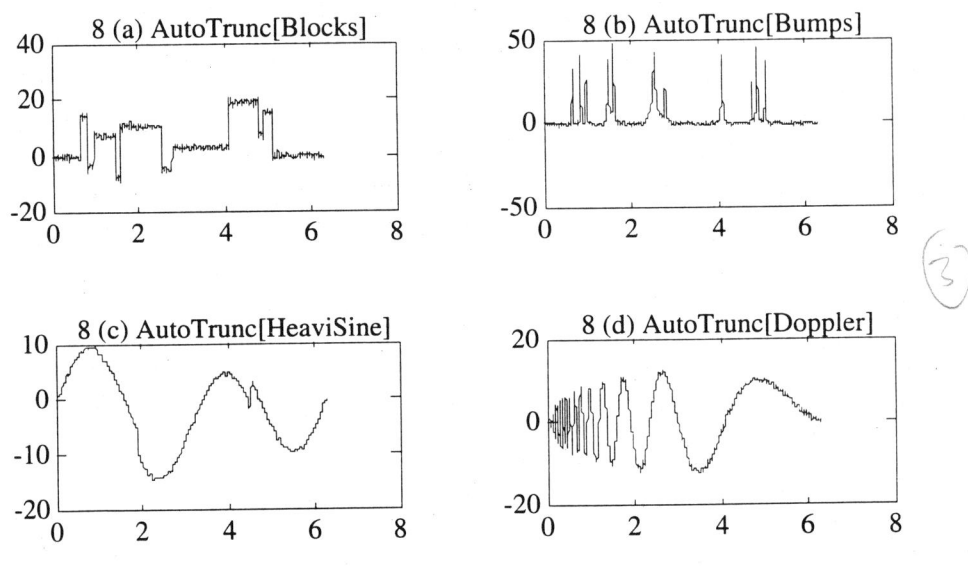

FIGURE 8.

3. Why it works.

3.1. Why it works: Data compression. A depiction of the wavelet shrinkage method in operation is given in Figure 9. Here we use Haar-basis shrinkage on a noisy version of object *Blocks*. The figure shows the original, noisy data (a), the noisy Haar coefficients (c), the thresholded coefficients (d), and the reconstruction (b). The method works because the Haar transform of the noiseless object *Blocks* compresses the ℓ^2 energy of the signal into a very small number of (consequently) very large coefficients. On the other hand, Gaussian white noise in any one orthogonal basis is again a white noise in any other (and with the same amplitude). Thus, in the Haar basis, the few nonzero signal coefficients really stick up above the noise. Therefore, the thresholding has the effect that it kills the noise while not killing the signal.

For a more formal argument, suppose we have data $d_i = \theta_i + \epsilon z_i$, $i = 1, ..., n$, where z_i is a standard white noise, and we wish to recover (θ_i). The ideal diagonal projector is the one which "keeps" all coefficients where θ_i is larger in

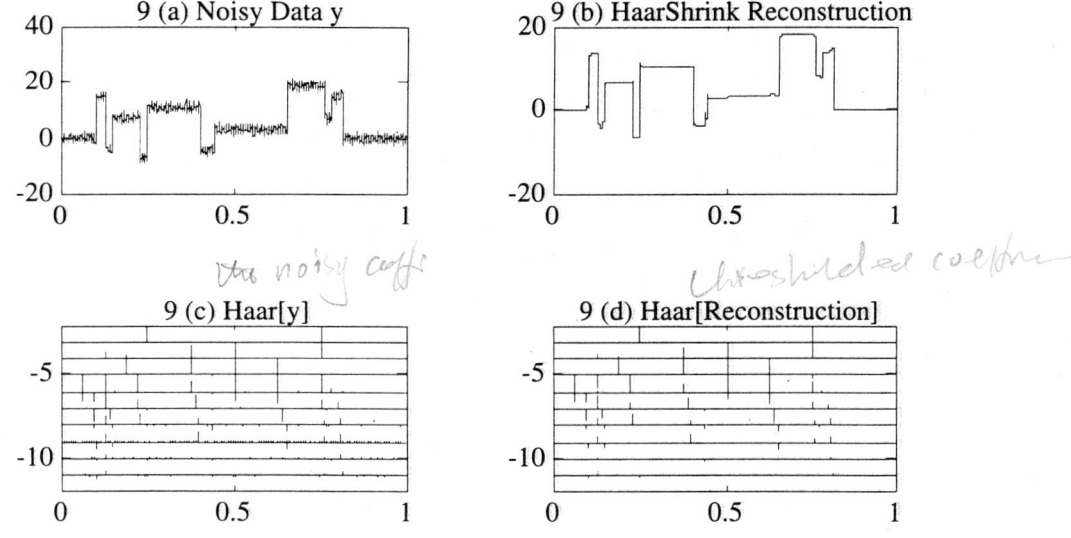

FIGURE 9.

amplitude than ϵ, and "kills" all coefficients where θ_i is smaller in amplitude than ϵ. (This ideal is unattainable, since it requires knowledge of θ, which we don't know). The ideal mean squared error is

$$R(\hat{\theta}^{IDEAL}, \theta) = \sum_i \min(\theta_i^2, \epsilon^2).$$

Define the "compression number" c_n as follows. With $|\theta|_{(k)} = k$-th largest amplitude in vector (θ_i), set $c_n \equiv \sum_{k>n} |\theta|_{(k)}^2$. This is a measure of how well the vector θ can approximated by a vector with n nonzero entries. Setting $N(\epsilon) = \#\{i : |\theta_i| \geq \epsilon\}$,

$$\sum_i \min(\theta_i^2, \epsilon^2) = \epsilon^2 \cdot \#\{i : |\theta_i| \geq \epsilon\}$$

$$+ \sum_i \theta_i^2 1_{\{i:|\theta_i|\leq\epsilon\}} = \epsilon^2 \cdot N(\epsilon) + c_{N(\epsilon)},$$

so this ideal risk is explicitly a measure of the extent to which the energy is compressed into a few big coefficients. (For more on this connection, see [**9**].)

Figure 10 shows the extent to which the different orthogonal bases compress the objects. The logarithm of the compression numbers is shown, plotted against n. The medium heavy line shows compression numbers in the Fourier Basis; the very heavy line (consisting of very closely spaced '+' signs – see 10(a)) marks the Haar compression numbers; and the thin line marks the compression using nearly-symmetric Daubechies wavelets having 8 vanishing moments. The wavelet basis generally wins, though with object *Blocks*, the Haar basis wins. Hence, ideal diagonal projectors work better in the wavelet basis than in the

Fourier basis. Traditional methods of smoothing are effectively little else than

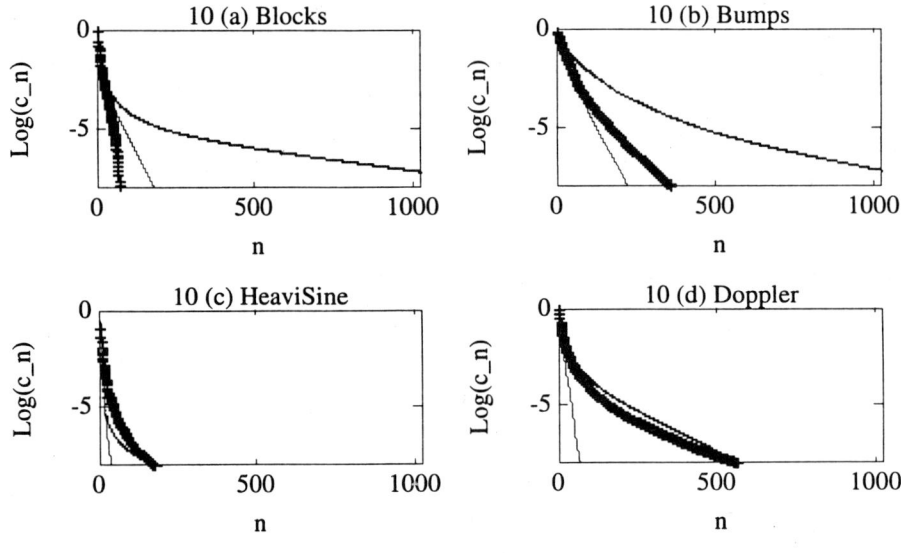

FIGURE 10.

(non-ideal) diagonal projectors in the Fourier basis. At the same time, soft-thresholding closely mimics an ideal diagonal projector in the wavelet basis [12]. The compression advantages of the wavelet basis are responsible for the mean-squared error advantages of wavelet shrinkage.

3.2. Why it works II: Unconditional basis. A very special feature of wavelet bases is that they serve as unconditional bases, not just of L^2, but of a wide range of smoothness spaces, including Sobolev and Hölder classes.

As a consequence, "shrinking" the coefficients of an object towards zero, as with soft-thresholding, acts as a "smoothing operation" in any of a wide range of smoothness measures.

The same can not be said of the Fourier basis. Kahane, Katznelson, and De Leeuw (see reference in Y. Meyer's book, volume 1, page 1) have shown that for functions on the circle, given any sequence of Fourier coefficients in ℓ^2 – perhaps the coefficients of an object that has square-integrable singularities on a countable dense subset of the circle – there is a continuous function that has each one of its Fourier coefficients *larger* than the given coefficients. In other words, the *smaller* coefficients correspond to the more bizarre object.

This can all be illustrated by example on the computer. In Figure 11 we display two signals – one a signal gathered by a seismic exploration crew, another an NMR spectrum. We also display reconstructions using only the 100 largest coefficients in the wavelet domain and in the Fourier domain, respectively. Note that reconstructions from thresholding in the Fourier domain display a kind of

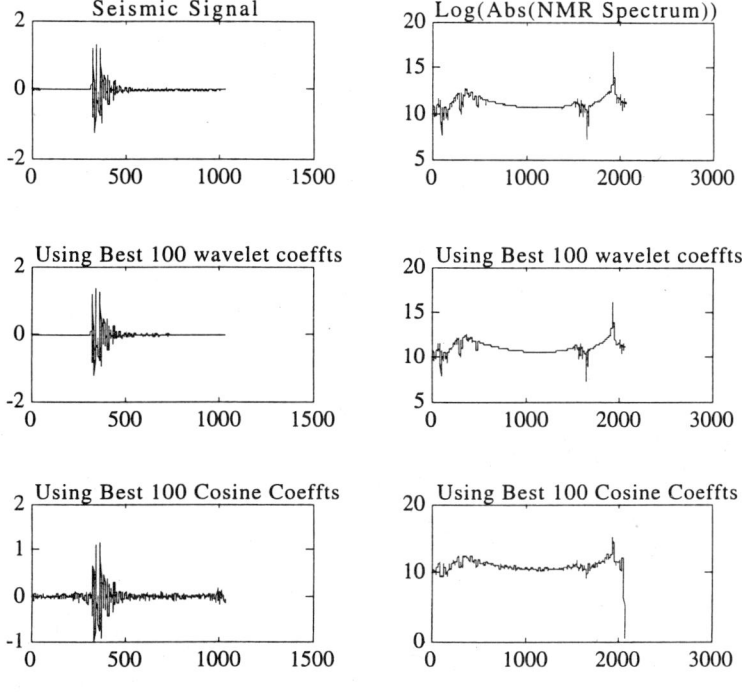

FIGURE 11.

"noise" throughout the signal – a very bad pointwise behavior. This illustrates that simple thresholding in the Fourier domain will not only give worse mean-squared errors, it will give unacceptable visual artifacts as well.

We conclude that the shrinkage of coefficients by soft-thresholding is in some sense visually adapted to use with the wavelet transform. Compare also [10].

4. Extensions: images, photon counts, densities, spectra.

The de-noising method of section 2 applies surprisingly widely. For example, if we had two-dimensional image data $y_{i_1, i_2} = f(i_1/m, i_2/m) + \epsilon z_{(i_1, i_2)}$ $i_1, i_2 = 0, ..., m - 1$ with $z_{(i_1, i_2)}$ white Gaussian noise, we would just use a 2-d pyramid filtering, and proceed as before, using the same three-step formalism with $n = m^2$. Figure 12 presents a 2-d image de-noising example.

(The application of thresholding to 2-d wavelet-like transforms has been discussed by Simoncelli, et al.[28] and by DeVore and Lucier [6]. DeVore and Lucier even find, by a different route, thresholds of the same general form we suggest;

they propose $C\sqrt{\log(n)}\sigma$. Mallat's multiscale-edge denoising involves a related thresholding principle, compare [**25**].)

In low lighting, the photon counting model $N_{i_1,i_2} \sim Poisson(f(i_1/m, i_2/m))$ is appropriate. To such data we would apply the Anscombe (1948) variance-stabilizing transformation

$$y_{i_1,i_2} = 2 \cdot \sqrt{N_{i_1,i_2} + 3/8}, \quad i_1, i_2 = 0, ..., m-1$$

and act as if the data arose from the Gaussian white noise model, with $\sigma = 1$.

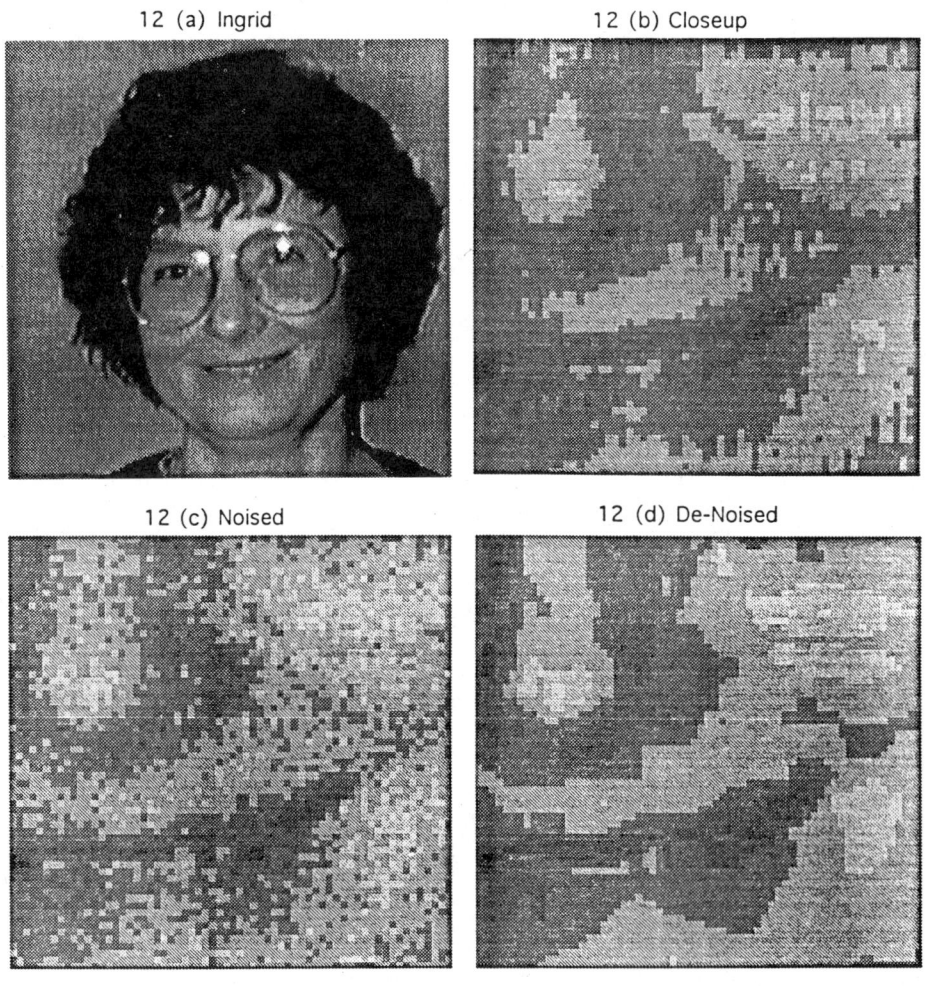

12 (a) Ingrid   12 (b) Closeup

12 (c) Noised   12 (d) De-Noised

FIGURE 12.

The results of doing this, in 1-dimension, on an ESCA spectrum, are shown in Figure 13.

Similarly, suppose we have a random sample $X_1, \ldots, X_m$, iid f, where f is an unknown density on $[0, 1]$. Partition $[0, 1]$ into $n = 2^{J+1}$ intervals, where $n \approx m/4$, and let N_i be the count of observations falling into the i-th interval. Then set

$$y_i = 2 \cdot \sqrt{N_i + 3/8}, \quad i = 1, \ldots, n,$$

and behave as if the y_i were Gaussian with mean $2 \cdot \sqrt{f(i/n)}$ and variance 1. This is connected with John Tukey's "Rootogram".

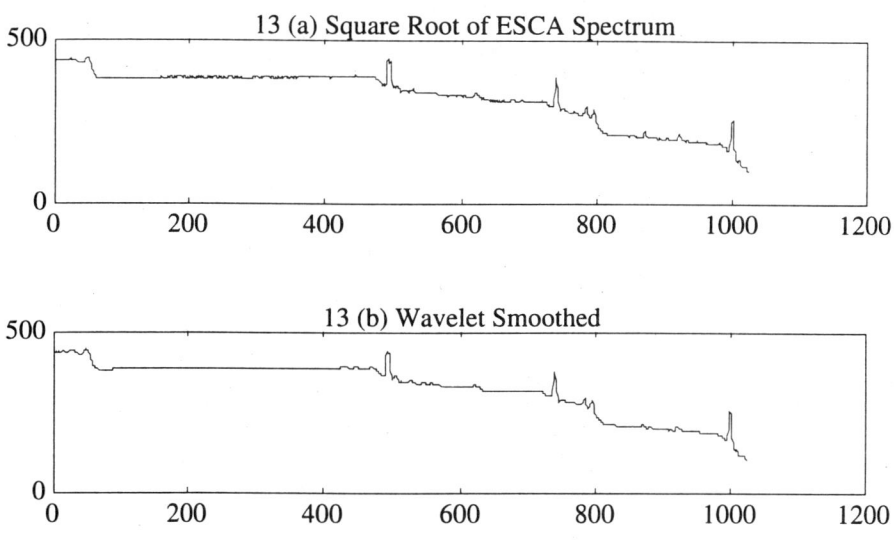

Data Courtesy of Jean-Paul Bibérian (Marseille)

FIGURE 13.

In another direction, suppose we have time series data $(x_t)_{t=0}^{n-1}$, $n = 2^{J+1}$, and we wish to estimate the spectral density function $f(\xi)$ of the (supposed) underlying second-order stationary process. We calculate the periodogram

$$I_k = n^{-2} \left| \sum_t x_t e^{i2\pi(t-1)(k-1)/n} \right|^2, \qquad k = 0, \ldots, n-1,$$

and apply the Wahba (1980) variance-stabilizing transformation to the log-periodogram:

$$y_k = (\log(I_k) + \gamma) \cdot \frac{\sqrt{6}}{\pi}, \qquad k = 1, \ldots, n/2 - 1, n/2 + 1, \ldots, n-1$$

where $\gamma \sim .57721\ldots$ is the Euler-Mascheroni constant, and a modification is required for the exceptional Fourier frequencies $k = 0, n/2$. This object might be called the "Log-o-Gram". We then treat the y_k as if they were Gaussian white noise data, with mean $\log(f(\xi_k)) \cdot \frac{\sqrt{6}}{\pi}$ and variance 1; here $\xi_k = 2\pi k/n$. The results of doing this for an AR(6) process which has roots near the unit circle are indicated in Figure 14. This general approach to time series spectra has been investigated by Hong-Ye Gao in his Berkeley Ph.D. thesis. Independently, P. Moulin [**27**] has suggested an approach based on this idea, and proposed a number of variations on choice of threshold, and has generalized the approach to the study of problems in radar imaging.

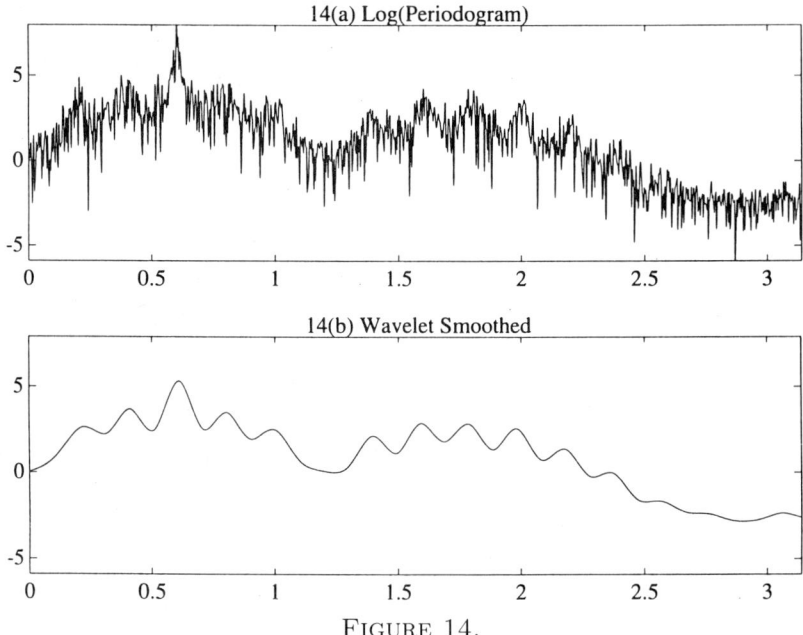

FIGURE 14.

In this rapid tour, we are cutting a few corners. A careful analysis of the theory underlying the Gaussian white noise model shows that for treating the density and spectral density case, we ought to use resolution-dependent thresholds which depend on the large-deviations properties of Poisson and Exponential noise. Otherwise we will tend to see tiny noise-induced 'blips' in an otherwise smooth curve (compare Figure 14). Hong-Ye Gao is writing his Ph.D. thesis at Berkeley in part on a finer analysis of this question in the time series setting; Eric Kolaczyk is writing his Ph.D. thesis at Stanford in part on an analysis of this question in the density setting.

5. Discrete inverse problems.

Many interesting problems having to do with noisy data involve *indirect* measurements. Here we obtain measurements

$$y_i = (Kf)(t_i) + \epsilon z_i$$

where Kf is a transformation of f. Examples include: Fourier transformation (magnetic resonance imaging), Laplace transformation (fluorescence spectroscopy), Radon Transformation (medical imaging) and many convolutional transformations (gravity anomalies, infrared spectroscopy, extragalactic astronomy).

Luckily, wavelet methods extend to handle various inverse problems as well. In some sense, such problems become problems of recovering wavelet coefficients in the presence of *non-white* noise. I will briefly discuss two simple examples.

5.1. Numerical differencing. Suppose we wish to reconstruct the discrete signal $(x_i)_{i=0}^{n-1}$, but we have only noisy data about the cumulative of x:

$$d_i = \left(\sum_{t=0}^{i} x_t \right) + \sigma z_i, \ldots i = 1, \ldots, n,$$

where z_i is a standard white Gaussian noise. We may attempt to invert this relation, forming the differences

$$y_i = d_i - d_{i-1},$$

with $y_0 = d_0$, of course. This is equivalent to observing

$$y_i = x_i + \sigma \cdot (z_i - z_{i-1}).$$

i.e. observations in a non-white noise.

We propose to reconstruct (x_i) by a three-step process similar to section 2, only with a threshold that is *level-dependent*. We choose this threshold by the rule

$$t_{j,n} = \sqrt{2 \log(n)} \cdot (2\sigma)/\sqrt{n} \cdot 2^{(J-j)/2}, \qquad j = j_0, \ldots, J;$$

this gives the reconstruction depicted in Figure 15. The situation in the wavelet domain is depicted in Figure 16. Note that the threshold is much larger at high-resolution levels than at low ones.

This scheme for thresholding may be motivated as follows. The noise in the wavelet transform is, at each resolution level, a Gaussian noise which is approximately stationary. The variance of the noise at level j grows roughly like 2^j (this is visually apparent). With this resolution-dependent thresholding, the noise is heavily damped, while the main structure in object "Bumps" persists. If we try traditional approaches instead, we get the results in Figure 17. Ideal Fourier damping is unable to suppress the noise.

5.2. Discrete-time deconvolution. Suppose we wish to reconstruct the discrete signal $(x_i)_{i=0}^{n-1}$, but we have only noisy data about a blurred-out x:

$$d_i = (k \star x)_i + \sigma z_i, \qquad i = 1, \ldots, n,$$

where $k \star x$ denotes a discrete convolution $\sum_u k_u x_{t-u}$ and z_i is a standard white Gaussian noise. (We cut corners by ignoring edge-effects.) Assume that we have

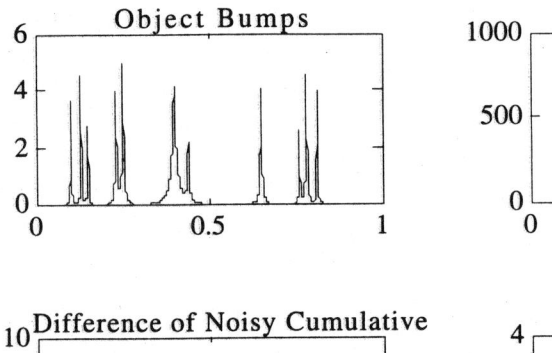
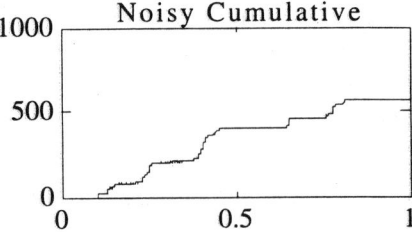
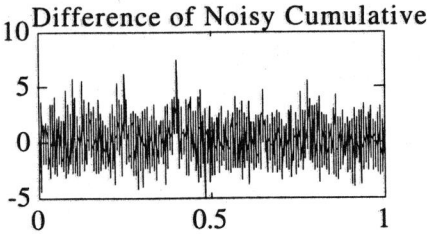
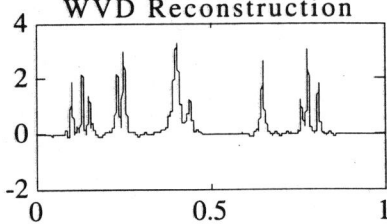

FIGURE 15.

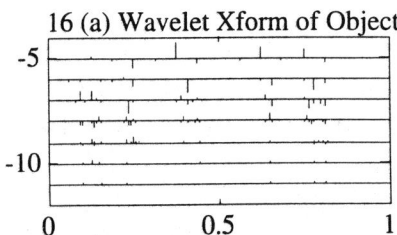
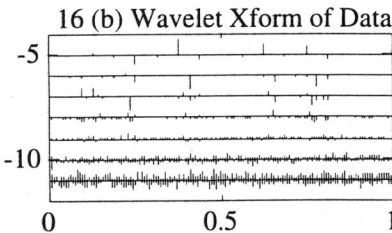

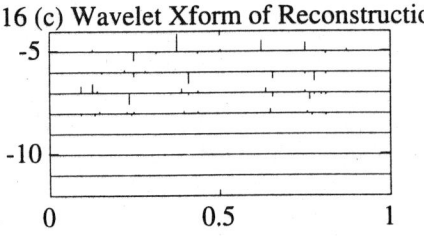

FIGURE 16.

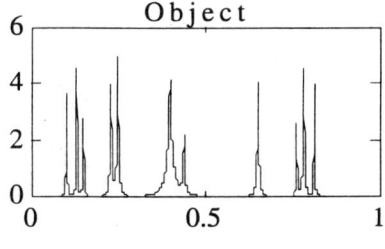

Object

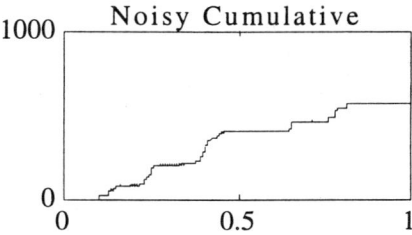

Noisy Cumulative

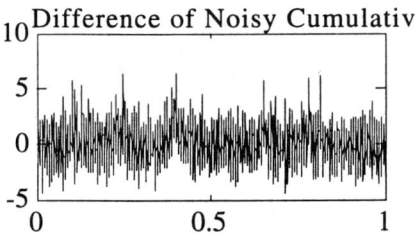

Difference of Noisy Cumulative

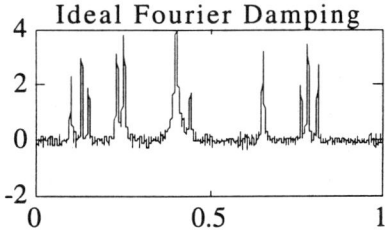

Ideal Fourier Damping

FIGURE 17.

a formal convolution inverse k^{-1}; we may attempt to invert this relation, forming

$$y_i = (k^{-1} \star d)_i.$$

This is equivalent to observing

$$y_i = x_i + \sigma \cdot (k^{-1} \star z)_i;$$

i.e. observations in a non-white noise.

We propose to reconstruct (x_i) by a three-step process similar to section 2, only again with a threshold that is *level-dependent*. We choose this threshold by the rule

$$t_{j,n} = \sqrt{2 \log(n)} \cdot \mathrm{MAD}((w_{j,k})_k)/.6745,$$

where $\mathrm{MAD}((v_i)_i) = \mathrm{Median}((|v_i|)_i)$.

We apply this idea to the system where k is a finite length recursive filter and k^{-1} a finite-length moving average $(1, -1.8, .81)$. This gives the reconstruction that was depicted earlier in Figure 2. The situation in the wavelet domain is depicted in Figure 18. Note that the threshold is again much larger at high-resolution levels than at low ones.

The motivation for this thresholding scheme is similar to that in §5.1. The noise in the wavelet transform is, at each resolution level, a Gaussian noise which is again approximately stationary. We now estimate the variance of the noise by assuming that "most" of the empirical wavelet coefficients at each resolution level are noise, and hence that the median absolute deviation reflects the size of the typical noise. The MAD/.6745 is an estimate of the noise standard deviation.

18 (a) Wavelet Xform of Object

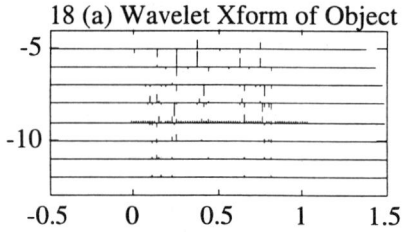

18 (b) Wavelet Xform of Data

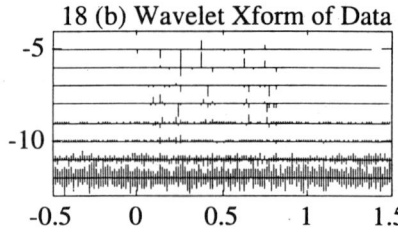

18 (c) Wavelet Xform of Reconstruction

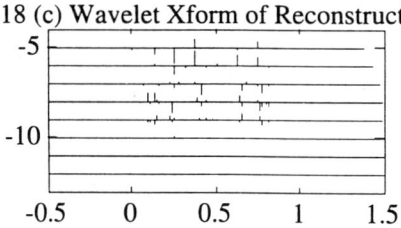

FIGURE 18.

With this resolution-dependent thresholding, the noise is heavily damped, while the main structure in object *Bumps* persists.

6. Continuous inverse problems.

General inverse problems can be conceptualized as observations

$$(6.1) \qquad d(t) = (Kf)(t) + z(t), \qquad t \in \mathcal{T}$$

where the index set might even be continuous. Mimicking section 5, we would ideally like to have an operator K^{-1} such that

$$(6.2) \qquad y(\xi) = (K^{-1}d)(\xi)$$

satisfies

$$(6.3) \qquad y(\xi) = f(\xi) + \tilde{z}(\xi)$$

where $\tilde{z} = K^{-1}z$ is a non-white noise. Unfortunately, in all the really interesting cases K^{-1} does not exist as a bounded operator on spaces to which the noise belongs.

We seek instead to mimick (6.2)–(6.3) in the wavelet domain. We want functionals $c_{j,k}$ with the property that

$$(6.4) \qquad c_{j,k}(Kf) = \langle \psi_{j,k}, f \rangle;$$

in other words, the linear functional $c_{j,k}$ applied to noiseless data gives the corresponding wavelet coefficient of f. Then applying these to noisy data

$$(6.5) \qquad\qquad y_{j,k} = c_{j,k}(d),$$

gives noisy measurements of the wavelet coefficients

$$(6.6) \qquad\qquad y_{j,k} = \langle \psi_{j,k}, f \rangle + \tilde{z}_{j,k}$$

where $\tilde{z}_{j,k}$ is an induced noise process. (6.5)–(6.6) make much better sense than (6.2)–(6.3), and one can follow the three-step de-noising procedure of section 2, using the MAD idea to obtain resolution-level dependent thresholds. This gives a practical method for dealing with rather general inverse problems.

When we apply this formalism to the Radon transform, the results are interesting. A two-dimensional tensor product wavelet basis has indices j, $k = (k_x, k_y)$, and also a directional preference $\epsilon \in \{1, 2, 3\}$. The functionals that solve the quadrature problem

$$c_{j,k}^{(\epsilon)}(Kf) = \langle \psi_{j,k}^{(\epsilon)}, f \rangle$$

have Riesz representers. To describe these, recall the set-up of the tomography problem. We have data

$$d(u, \theta) = (P_\theta f)(u) + z(u, \theta)$$

where $\theta \in [0, 2\pi]$ has to do with the projection angle, and $u \in \mathbf{R}$ with the foot of the projection ray. The representers $\gamma_{(j,k,\epsilon)}$ of the $c_{j,k}^{(\epsilon)}$ have the form

$$\gamma_{(j,k,\epsilon)}(u, \theta) = 2^j \cdot \gamma_{(0,0,\epsilon)}(2^j u - \cos(\theta)k_x - \sin(\theta)k_y).$$

The $\gamma_{(j,k,\epsilon)}$ are all "twisted" dilations of three fixed "mother representers" $\gamma_{(0,0,\epsilon)}$. As j increases, they concentrate around certain sine-curves in the (u, θ) plane. These sine-curves $2^{-j}(\cos(\theta)k_x - \sin(\theta)k_y)$ name certain positions $2^{-j}(k_x, k_y)$ in the original image space.

Figure 19 shows the three mother representers, and an example of a twisted dilation. The diagonal in the direction East-NorthEast is θ, the one in direction North-NorthWest is u. The directional sensitivity of the original wavelets is responsible for the fact that the representers effectively vanish for certain ranges of θ.

[7] develops a general formalism for addressing inverse problems using wavelets which generates the above examples as special cases. The idea is to develop a decomposition of the forward operator K in terms of wavelets and *vaguelettes* which, at a formal level, resembles the Singular Value Decomposition (SVD), but which uses a wavelet basis instead of an eigenfunction basis. The idea is that an eigenfunction basis, like the Fourier basis, will have trouble representing objects with spatial variability, and therefore a *Wavelet-Vaguelette decomposition* (WVD) will be a better way to represent many problems than the SVD.

19 (a) Gamma(0,0,Vert)

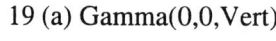
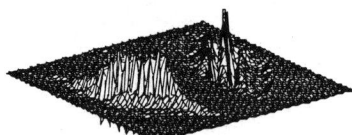

19 (b) Gamma(0,0,Horiz)

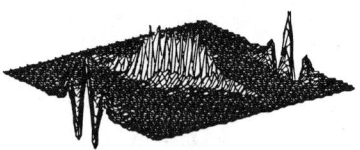

19 (c) Gamma(0,0,Diag)

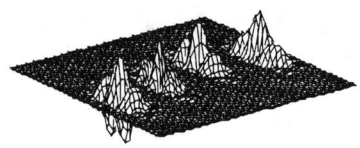

19 (d) Gamma(j=3,(kx=4,ky=2),Diag)

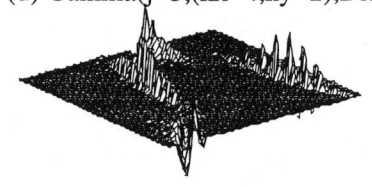

FIGURE 19.

The WVD starts from the representers $\gamma_{j,k}$ solving the quadrature relations (6.4) and identifies constants κ_j so that the functions

$$u_{j,k} = \gamma_{j,k} \cdot \kappa_j$$

make a set of functions with norms bounded above and below. Then the functions $v_{j,k} = K\psi_{j,k}/\kappa_j$ are biorthogonal to $u_{j,k}$ in the data space:

$$[u_{j,k}, v_{j',k'}] = \delta_{j,k;j',k'}.$$

Next one checks that the two sets $(u_{j,k})$ and $(v_{j,k})$ are almost-orthogonal, in the sense that

$$\left\| \sum a_{j,k} u_{j,k} \right\|_{L^2(dt)} \asymp \|(a_{j,k})\|_{\ell^2} \asymp \left\| \sum a_{j,k} v_{j,k} \right\|_{L^2(dt)}.$$

It results that the formal relations

$$(6.7) \qquad\qquad Kf = \sum [Kf, u_{j,k}]\kappa_j v_{j,k}$$

and

$$(6.8) \qquad\qquad f = \sum [Kf, u_{j,k}]\kappa_j^{-1}\psi_{j,k}$$

have a content which can be made rigorous. When this is so, inversion from noisy data may be defined by soft thresholding

$$\hat{f} = \sum \eta_{t_j}([y, u_{j,k}])\kappa_j^{-1}\psi_{j,k}$$

with threshold

$$t_j = \sqrt{2log(2^j)}\,\widehat{\mathrm{SDEV}}([y, u_{j,k}])$$

which is an abstract generalization of the earlier examples.

To understand when this all works, compare (6.7)–(6.8) with the usual SVD relations

$$Kf = \sum [Kf, f_\nu]\lambda_\nu f_\nu$$

and

$$f = \sum [Kf, f_\nu]\lambda_\nu^{-1} e_\nu;$$

here the e_ν are eigenfunctions of the operator $K^\star K$ and $f_\nu = Ke_\nu/\|Ke_\nu\|$.

In some sense the approach works when wavelets are "almost eigenfunctions" of $K^\star K$. That is, when the WVD may be defined, we have

$$K\psi_{j,k} = \kappa_j v_{j,k}; \qquad K^\star u_{j,k} = \kappa_j \psi_{j,k};$$

so K is mapping wavelets into vaguelettes and $K^\star$ is mapping vaguelettes into wavelets. Only special operators K will exhibit such character (in the same way that "only" Calderón-Zygmund operators map "atoms" into "molecules"). When one has such an operator, wavelets offer an almost-SVD, where we give up exact invariance under $K^\star K$ in order to get much better representation of the objects f of interest. Examples where the WVD may be defined include Radon transform, Fractional Integration, and various convolution operators.

7. What's so special about wavelets?

Many groups have independently developed methods for noise suppression which are also based on wavelet thresholding in some sense. I think here of Mallat and collaborators (Courant), Coifman and collaborators (Yale), and Healy and collaborators (Dartmouth). These other groups have found that wavelet thresholding methods work well in problems ranging from photographic image restoration to medical imaging. R.A. DeVore (South Carolina) and B.J. Lucier (Purdue) have also come to thresholding, motivated by approximation-theoretic arguments. P. Moulin of Bell Labs has introduced wavelet thresholding techniques for radar imaging.

This agreement of diverse empirical, engineering, and mathematical work is very encouraging, and suggests that wavelet shrinkage will soon have a large impact on how scientists treat noisy data. There is also theoretical work in mathematical statistics, which we describe in a moment, which "proves" that wavelet shrinkage offers special properties.

7.1. What's so special about wavelets I: Advantages in statistical theory. Wavelet shrinkage possesses a disarming simplicity. In fact it achieves many theoretical goals simultaneously. For example, in the context of section 2, [14] shows that in estimating a function of unknown Hölder smoothness at

a point, the estimator $\hat{f}(t_0)$ attains within a constant factor the minimax behavior among all measurable procedures. Formally this goes as follows. Let $\mathcal{F}(C,\alpha)$ denote the collection of all functions Hölder(-Zygmund) continuous with exponent α and Hölder seminorm bounded by C. For $0 < \alpha < 1$ this means $|f(x) - f(y)| \leq C|x - y|^\alpha$, with obvious extensions to $\alpha \geq 1$. Then

$$\sup_{\mathcal{F}(C,\alpha)} E(\hat{f}^*(t_0) - f(t_0))^2 \leq Const \cdot \log(n)^r \cdot \inf_{\hat{f}} \sup_{\mathcal{F}(C,\alpha)} E(\hat{f}(t_0) - f(t_0))^2,$$

where $r = 2\alpha/(2\alpha + 1)$, valid for $0 < C < \infty$, and for $0 < \alpha < \alpha_0$, where α_0 is set by the regularity of the underlying wavelets. Hence a single estimator is within a logarithmic factor of minimax over every Hölder ball. Recent results in statistical decision theory due to Lepskii and to Brown and Low show that this logarithmic factor cannot be removed. No estimator can do essentially better than this uniformly over such a broad range of balls.

[10] shows that the same estimator attains, within logarithmic factors, the optimal rate of convergence in a global ℓ^2 norm simultaneously over all Besov and Triebel balls in a certain range; this range is limited by the wavelet employed. If now $\mathcal{F}(C)$ denotes a Besov ball $B^\sigma_{p,q}(C)$ with smoothness degree σ obeying $1/p < \sigma < R$, where R is the regularity of the wavelet employed, then

$$\sup_{B^\sigma_{p,q}(C)} E \sum_i (\hat{f}^*(t_i) - f(t_i))^2 \leq Const \cdot \log(n)^r \cdot \inf_{\hat{f}} \sup_{B^\sigma_{p,q}(C)} E \sum_i (\hat{f}(t_i) - f(t_i))^2,$$

where $r = 2\sigma/(2\sigma + 1)$.

[13] shows that there is a way to remove these logarithmic factors by clever choice of thresholds; [15] shows how to use Stein's Unbiased estimate of Risk [29] in order to do so in a practical way.

[16] shows that the same conclusions hold in a wide variety of norms in the Besov and Triebel scales; it is not necessary to use L^2-type losses.

In [23], [22], and [17], Johnstone, Kerkyacharian, and Picard have discovered a variety of nice properties of wavelet shrinkage in the density model, $X_1, \ldots, X_n$ i.i.d. f, though not with the estimator described above, and by completely different methods of proof.

These properties are unprecedented in several ways. For many years, statisticians in the USA. Europe, and Russia have developed techniques for smoothing noisy data for the purpose of signal extraction. Typically, they were working with convolutional smoothers, stiffness-penalized splines, or Fourier-domain damping, and so the questions of how much to smooth, penalize, or damp were paramount [30]. Wavelet shrinkage completely avoids these issues, is much simpler, and has very broad near-optimality properties never dreamed of before, and not attainable by older methods. The method achieves, within a logarithmic factor, the minimax risk over each functional class in a wide variety of smoothness classes and with respect to a wide variety of losses [16]. Older methods achieve the minimax rate only over special subsets of the full range of Besov and Triebel classes [**14**].

Traditional methods, except for the "amount of smoothing" issue, are linear, and cannot compete effectively with the wavelet shrinkage method in cases of high spatial variability – either in practice (e.g. Figures 6, 7, 8) or in theory. In estimating functions of bounded variation, linear methods cannot attain the optimal rate, nor can methods with ideal choice of "amount to smooth"; the wavelet shrinkage method of section 2 attains a mean-squared error of size $(log(n)/n)^{2/3}$ based on n observations, while linear and adaptive linear methods attain only an error of size $n^{-1/2}$.

For inverse problems, WVD has parallel optimality properties. An example of its quantitative advantages is the ability to recover objects in the 2-dimensional Bump Algebra from Radon data, with an error of order $n^{-4/7}$ from n samples, while the SVD and traditional linear methods only achieve the rate $n^{-2/5}$; see [**7**]. Presumably this means that filtered backprojection and similar linear methods now employed in medical scanners can be outperformed by wavelet shrinkage, when the object to be recovered is spatially variable – possessing edges and highly localized features.

7.2. What's so special about wavelets II: Mathematical properties. An interesting aspect of the above theorems in mathematical statistics is how they rely on fundamental facts about wavelet bases derived by mathematicians for other purposes. For the reader's convenience, we briefly point out that various mathematical results on the special properties of wavelet bases imply corresponding statistical results just stated.

Unconditional basis property. A primary preoccupation of Meyer-Lemarie and Frazier and Jawerth has been in showing that wavelets offer unconditional bases of L^2 and also of many smoothness spaces as well. As we have seen, this property of being simultaneously an unconditional basis of many spaces means that shrinkage of wavelet coefficients is a smoothing operation in many different norms simultaneously. [**9**] and [**16**] have shown how this property leads explicitly to the near-minimaxity results quoted above.

Spatial adaptation property. A primary preoccupation of Ron DeVore has been to show that wavelets are good at representing objects in certain special Besov spaces B_τ. The statistical implication: wavelet shrinkage has therefore at least the same ability to estimate spatially adaptive phenomena as various adaptive partitioning and variable-bandwidth kernel estimation schemes in common use in statistics, and conjectured to have good behavior, but for which rigorous theory is harder to get than for wavelets [**12**].

Almost-diagonality property. A considerable body of research by French waveleticians Yves Meyer, Stephane Jaffard, Philippe Tchamitchian, and U.S. waveleticians Michael Frazier and Björn Jawerth has been to show that many important mathematical operators (e.g. certain convolutional operators) are almost diagonal in a wavelet basis. The implication for us is that when we need to solve an inverse problem involving such an operator, wavelets are almost as good as eigenfunctions at representing the operator under study, but in our applications they typically are far better than eigenfunctions at representing the object to be recovered, hence the WVD approach beats the traditional SVD approach.

8. Beyond wavelets.

So far, we have emphasized the use of wavelet bases, and the development of methods which are simple, yet in some sense provably optimal for use in those bases. These developments prove that wavelets solve theoretical problems which had attracted considerable activity over many years, and which resisted solution by non-wavelet techniques.

Now we turn to applications which are more complicated and for which the theory is not yet complete. The applications involve modifying or extending wavelets in various ways. Here we present computational examples indicating some of the motivation and some of the possibilities.

8.1. Beyond wavelets I: Adaptive choice of basis.
As indicated in the introduction, Coifman, Meyer and Wickerhauser have introduced a family of orthogonal bases, of which wavelet and Fourier bases are special cases, for which there exist fast transforms and which offer promise for analysis of signals which exhibit moderate-duration oscillatory phenomena.

These ideas, translated into a statistical setting, pose a number of interesting issues. As indicated in section 3, if $\mathcal{B}$ denotes an orthogonal basis, the best mean-squared error among linear estimates diagonal in the basis $\mathcal{B}$ is essentially

$$R(f, \mathcal{B}) = \sum_i \min(\theta_i^2, \epsilon^2),$$

where ϵ^2 is the noise level. This goal is attainable only if we know a priori which coordinates are large and which are small, and hence only with the aid of a *coordinate-oracle*. We never have such an oracle at our disposal; nevertheless this is the goal we shall consider. We call it the "Ideal Risk", since it is the Risk attainable by an ideal oracle-assisted algorithm.

Now, when a whole collection of bases is available, it becomes of interest to consider what could be obtained with the aid of a *basis-oracle*, which presents us with the basis $\mathcal{B}^*$ obeying

$$R(f, \mathcal{B}^*) = \min_{\mathcal{B}} R(f, \mathcal{B}).$$

To what extent can we develop a method which approaches the ideal risk, i.e. the risk attainable with a coordinate-and-basis oracle?

Preliminary experiments with the use of Stein's Unbiased Estimate of Risk indicate that we can come surprisingly close to this ideal performance. As background, we recommend [12] where the concept of ideal risk is discussed more carefully, and precedents for nearly attaining it are given.

In [15], Donoho and Johnstone have introduced a method of selecting among a family of estimators for use in the wavelet basis – the selection is based on Stein's Unbiased Estimation of Risk (SURE). SURE had previously been used in selecting among families of linear estimates; [15] showed that it could be used in selecting among soft-thresholding estimates, achieving results nearly as good as could be obtained with an oracle.

In the basis selection problem, we fix the threshold $t_n = \sqrt{2\log_e(n\log_2(n))}$, and we define

$$\mathrm{SURE}(y, \mathcal{B}) = \epsilon^2 \cdot \left(n - 2\sum_i 1_{|y_i| \le t_n \epsilon} + \sum \min((y_i/\epsilon)^2, t_n^2) \right).$$

This has the property that when the model $y_i = \theta_i + \epsilon z_i$ holds in the basis $\mathcal{B}$, with z_i i.i.d. $N(0,1)$, then

$$E\,\mathrm{SURE}(y, \mathcal{B}) = E\|(\eta_{t_n\epsilon}(y_i)) - \theta_i\|_2^2.$$

This is true for all θ; it is the unbiasedness property which explains the initials "URE" in SURE. We therefore have tried selecting a basis by minimizing the SURE of the soft-thresholding estimator in that basis, i.e. finding a basis by the principle

$$\mathrm{SURE}(y, \hat{\mathcal{B}}) = \min_{\mathcal{B}} \mathrm{SURE}(y, \mathcal{B}).$$

It is easy to find a best basis by adapting the Best-Basis algorithm of Coifman and Wickerhauser to the SURE cost function.

To show how this works in examples, Figure 20 presents four signals – two highly oscillatory, where it is expected that wavelets will not work very well, and *Bumps* and *Doppler* from earlier examples, where we have already seen that wavelets work well. Figure 21 shows the same signals with noise. Figure 22 shows the same signals reconstructed using $\eta_{t_n\epsilon}$ in the Wavelet Packet basis selected by SURE. Note that the same software has been used, with the same settings on all four signals. The range of reconstructions is made possible by adaptive choice of basis.

For comparison, Figure 23 shows the results of reconstruction of the object *Mishmash* by Splines with automatically chosen tension parameter, by Wavelet Shrinkage, by Fourier Shrinkage, and by SURE selection of wavelet packet basis. Visually, at least, the adaptive basis method outperforms the more traditional techniques (if "wavelet shrinkage" can now be called traditional).

8.2. Beyond wavelets II: Segmented multiresolution analysis. One can view the "Best Basis" methods of Coifman, Meyer, and Wickerhauser as establishing an interesting paradigm with applications far beyond the wavelet-packets – cosine packets setting. For example, we describe work in progress on minimum entropy segmentation.

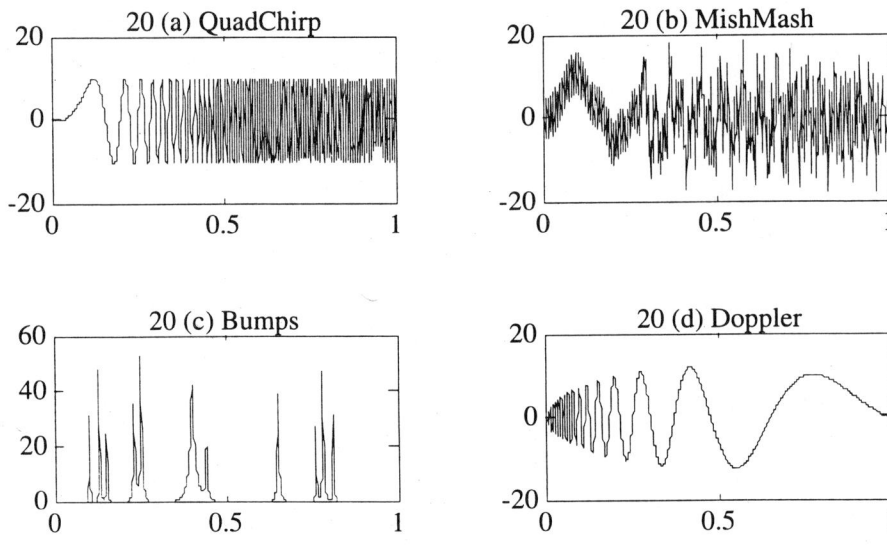

FIGURE 20.

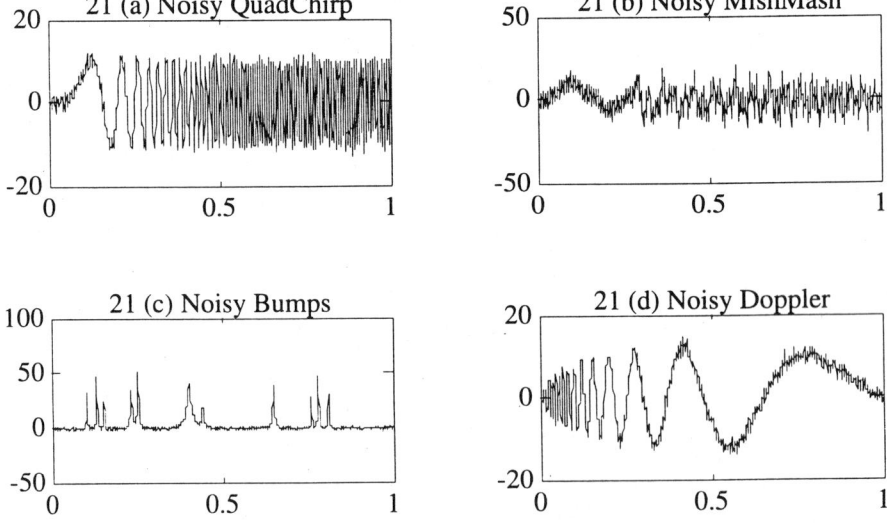

FIGURE 21.

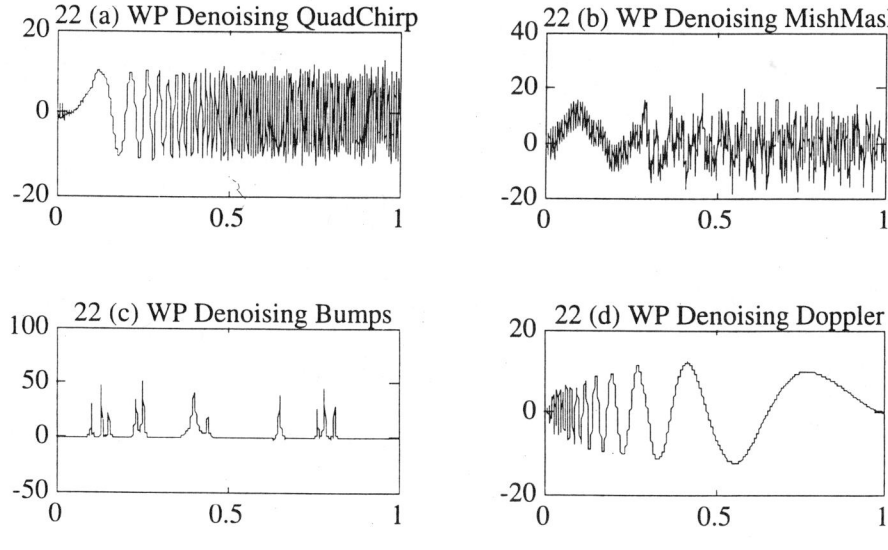

FIGURE 22.

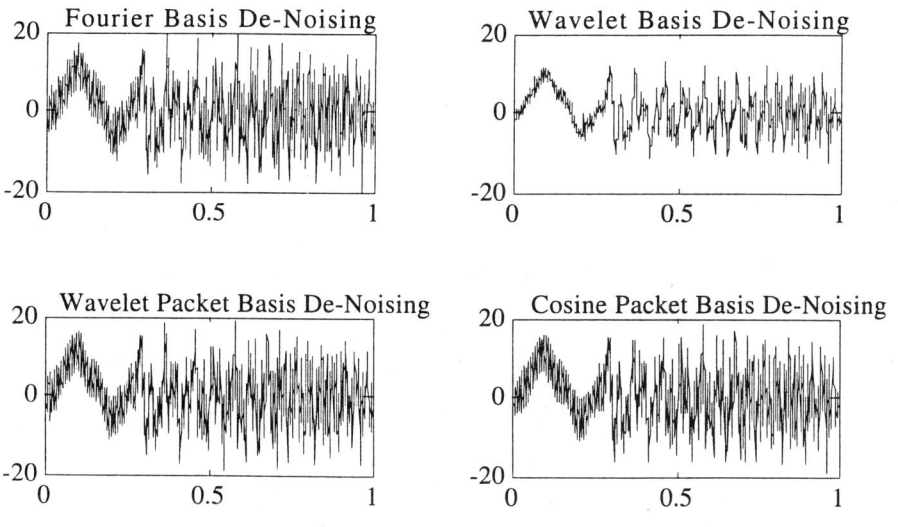

FIGURE 23.

For simplicity we discuss the 1-d case. A *segmented multiresolution analysis* V_j^t of $[0, 1]$ is a multiresolution analysis in which $[0, t]$ and $[t, 1]$ are somehow kept essentially separate, so that functions in V_j^t need not be continuous at the point t. (Compare the notions of splitting and merging in Andersson, Hall, Jawerth, and Peters [1], which could be used to implement segmented multi-resolution decomposition. The specific segmented multi-resolution we use is based on the average-interpolating multi-resolutions in [11]).

Obviously, in analyzing objects with discontinuities at the points t, a multiresolution analysis which permits discontinuities at the point t permits better approximation and, ultimately, better compression. Figure 24 makes this point. We have a piecewise linear function, together with a standard (average-interpolating) multi-resolution refinement, and a segmented refinement. The standard multi-resolution smooths out the edge; the segmented refinement preserves the edge.

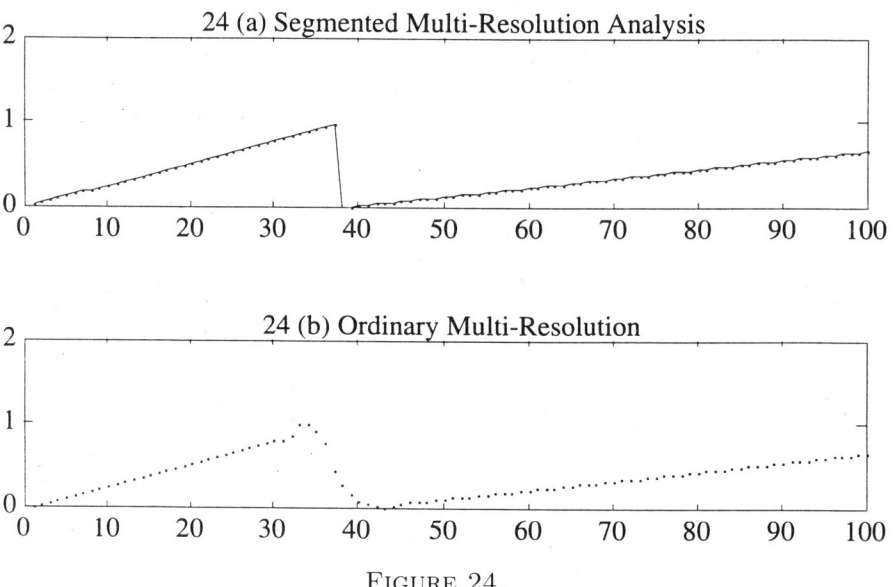

FIGURE 24.

Corresponding to the segmented refinement is a segmented wavelet transform, with coefficients $\alpha_{j,k}^t$, which depend on the segmentation point t. When we analyze a function π^t which is piecewise polynomial, a discontinuity being allowed only at t, then $P_j^t \pi^t = \pi^t$, which implies that the wavelet coefficients $\alpha_{j,k}^t$ are identically zero for such an object (when the parameter t of the transform and the site t of the discontinuity are really the same.) Figure 25 shows the behavior of de-noising applied to a segmented wavelet transform, and a comparison with de-noising of a non-segmented wavelet transform. The improvement in the neighborhood of the discontinuity is dramatic: a significant reduction in the Gibbs phenomena there.

Of course, in Figure 25 we are showing what happens when the wavelet transform is segmented exactly at the point of discontinuity. How are we to obtain, in analyzing noisy data, information about the proper segmentation point? Viewing the collection of segmented wavelet transforms with different values of t as a collection of bases $\mathcal{B}^{(t)}$, this is really a problem of selecting a best basis. Therefore we propose a *best-basis segmentation*

$$\mathrm{SURE}(y, \hat{t}) = \min_t \mathrm{SURE}(y, \mathcal{B}^{(t)}).$$

For the dataset in question, the visual performance of the resulting estimate is indistinguishable from that in Figure 25.

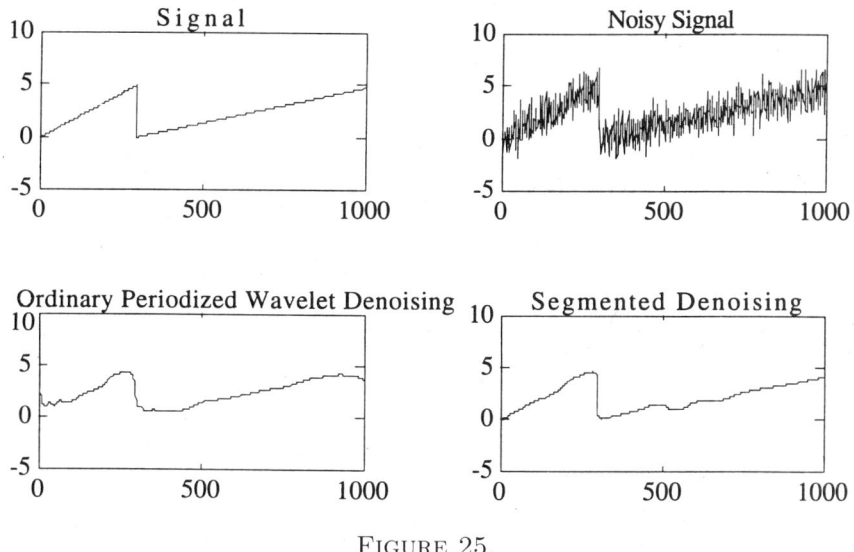

FIGURE 25.

8.3. Beyond wavelets III: Nonlinear multiresolution analysis. In our discussion so far, "noise" has always meant Gaussian noise, or something close to Gaussian so that Central Limit Theorem considerations apply. In some sense, linear, orthogonal wavelet analysis is naturally tied to an assumption of Gaussianity. If we instead have very non-Gaussian noise, then *nonlinear* multiresolution analysis becomes practically de rigueur. To see why, we consider Cauchy noise (the symmetric stable law of index 1), independent and identically distributed, with density $f(t) = \pi^{-1}(1 + t^2)^{-1}$. In Figure 26 we superpose such noise on a sine-curve. The occasional large noise spikes completely dominate the plot scaling, and nothing of interest remains visible. If we try to "smooth away" the noise by setting to zero the noisy wavelet coefficients at fine scales and inverting the wavelet transform, we still don't see anything useful (Figure 27).

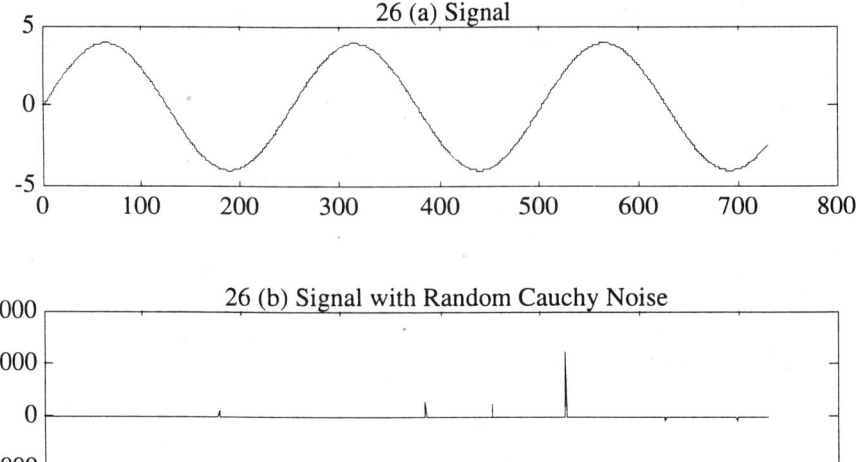

FIGURE 26.

Consider now a simple recursive nonlinear multiresolution scheme based on decimating by factors of 3. The fine-to-coarse mapping is obtained by grouping the signal in triplets of successive points, and replacing each group of three by a single number – the median of the group of 3. (This is a sort of nonlinear Haar analysis, since dyadic Haar wavelets correspond to grouping data in pairs and keeping only the mean of each pair.) This triadic nonlinear coarsening operator gives rise in an obvious way to a telescoping nonlinear multiresolution decomposition. Figure 28 shows the result of setting to zero the fine scale coefficients of this nonlinear triadic transform applied to the noisy data in Figure 26(b). This clearly performs much better than the linear recovery in Figure 27.

Theoretical work to date on nonlinear multiresolution analysis has been done by Ron DeVore (S. Carolina) and Bradley Lucier (Purdue). Interesting applied work with mammograms has been done by Rich Richardson (Univ. of Texas at San Antonio). Doug Martin and Andrew Bruce (Univ. Washington, Seattle), along with the author, have developed a variety of algorithms based on ideas from robust statistics.

Acknowledgements.

It is a pleasure to thank Iain Johnstone, Gérard Kerkyacharian, and Dominique Picard, with whom much of the theory described here has been developed, and to thank Yves Meyer, Ronald Coifman, and Ingrid Daubechies for encouragement at key moments. Specific inspirations provided by the work of Ronald DeVore and Björn Jawerth are also gratefully acknowledged, as well as stimulating conversations with Albert Cohen and Bradley Lucier. Thanks to

Andrew Bruce and Carl Taswell for many discussions about wavelet software. The NMR datasets were provided by Chris Raphael (Figure 1) and Jeff Hoch (Figure 11), the ESCA dataset by Jean-Paul Bibérian, the image dataset by Ingrid Daubechies, the seismic dataset by Paul Donoho. Many thanks to Tina Sharp for intense last-minute editorial work.

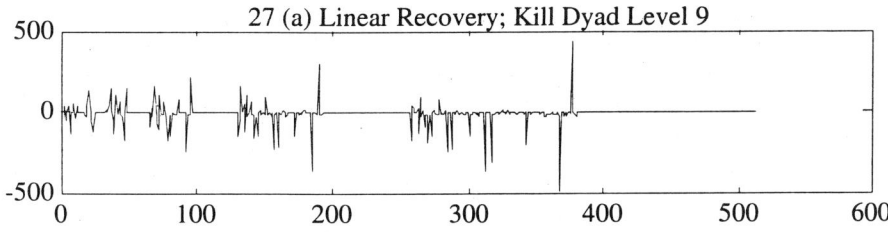

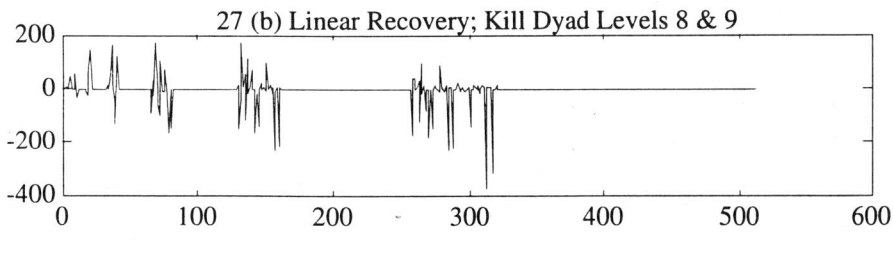

FIGURE 27.

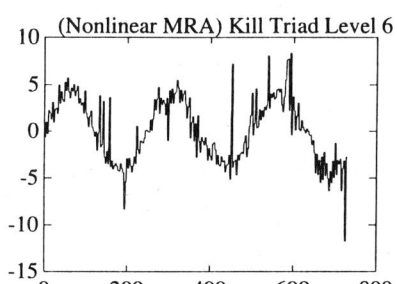

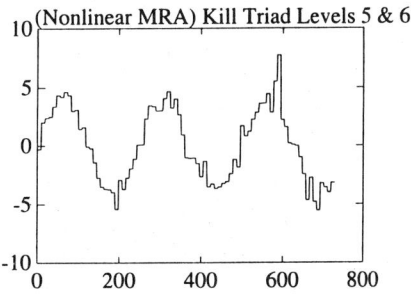

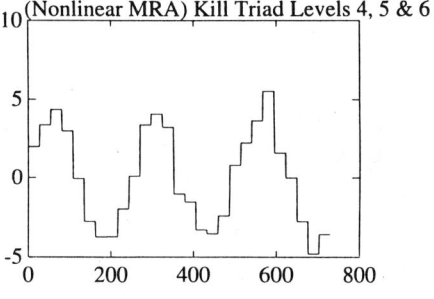

FIGURE 28.

References

1. L. Andersson, N. Hall, B. Jawerth and G. Peters, "Wavelets on closed subsets of the real line", in *Recent Advances in Wavelet Analysis*, Larry L. Schumaker and Glenn Webb (eds.), Academic Press, Boston, 1993.

2. A. Cohen, I. Daubechies, B. Jawerth and P. Vial, "Multiresolution analysis, wavelets, and fast algorithms on an interval", *Comptes Rendus Acad. Sci. Paris* (A) **316** (1992) 417-421.

3. R. R. Coifman and Y. Meyer, "Remarques sur l'analyse de Fourier à fenêtre", *Comptes Rendus Acad. Sci. Paris* (A) **312** (1991) 259-261.

4. R. R. Coifman, Y. Meyer and M. V. Wickerhauser, "Wavelet analysis and signal processing", pp. 153–178 in *Wavelets and Their Applications*, M. B. Ruskai et al. (eds.), Jones and Bartlett (Boston) 1992.

5. R. R. Coifman and M. V. Wickerhauser, "Entropy-based algorithms for best-basis selection", *IEEE Trans. Info. Theory* **38** (1992) 713-718.

6. R. A. DeVore and B. J. Lucier, "Fast wavelet techniques for near-optimal image processing", *Proc. IEEE Mil. Commun. Conf.*, Oct. 1992. IEEE Communications Society, NY, 1992.

7. D. L. Donoho, *De-Noising via Soft-Thresholding*, Tech. Rept., Statistics, Stanford, 1992.

8. D. L. Donoho, *Nonlinear solution of linear inverse problems by wavelet-vaguelette decomposition*, Tech. Rept., Statistics, Stanford, 1992.

9. D. L. Donoho, *Unconditional bases are optimal bases for data compression and for statistical estimation*, Tech. Rept., Statistics, Stanford, 1992.

10. D. L. Donoho and I. M. Johnstone, *Minimax risk over ℓ_p-balls*, Tech. Rept., Statistics, Univ. Calif., Berkeley, 1990.

11. D. Donoho, "Smooth wavelet decompositions with blocky coefficient kernels", to appear in *Advances in Wavelet Analysis*, L. L. Schumaker and G. Webb (eds.), Academic Press, Boston, 1993.

12. D. L. Donoho and I. M. Johnstone, *Ideal spatial adaptation via wavelet shrinkage*, Tech. Rept., Statistics, Stanford, 1992.

13. D. L. Donoho and I. M. Johnstone, *New minimax theorems, thresholding, and adaptation*, Tech. Rept., Statistics, Stanford, 1992.

14. D. L. Donoho and I. M. Johnstone, *Minimax estimation by wavelet shrinkage*, Tech. Rept., Statistics, Stanford, 1992.

15. D. L. Donoho and I. M. Johnstone, *Adapting to unknown smoothness by wavelet shrinkage*, Tech. Rept., Statistics, Stanford, 1992.

16. D. L. Donoho, I. M. Johnstone, G. Kerkyacharian and D. Picard, *Wavelet Shrinkage: Asymptopia?*, Tech. Rept., Statistics, Stanford, 1993.

17. D. L. Donoho, I. M. Johnstone, G. Kerkyacharian and D. Picard, *Density Estimation via Wavelet Shrinkage*, Tech. Rept., Statistics, Stanford, 1993.

18. B. Efron and C. Morris, "Data analysis using Stein's estimator and its generalizations", *J. Amer. Statist. Assn.* **70** (1975) 311–319.

19. J. Froment and S. Mallat, "Second-generation compact image coding with wavelets", in *Wavelets: a Tutorial in Theory and Applications*, C. Chui (ed.), Academic, Boston, 1992, 655-678.

20. Hong-ye Gao, *Choice of Thresholds for wavelet estimation of the log-spectrum*, Tech. Rept., Statistics, Stanford, 1993.

21. Hong-ye Gao, *Spectral Density Estimation via Wavelet Shrinkage*, Tech. Rept., Statistics, Stanford, 1993.

22. I. M. Johnstone, G. Kerkyacharian and D. Picard, "Estimation d'une densité de probabilité par méthode d'ondelettes", *Comptes Rendus Acad. Sciences Paris* (A) **315** (1992) 211-216.

23. G. Kerkyacharian and D. Picard, "Density estimation in Besov Spaces", *Statistics and Probability Letters* **13** (1992) 15-24.

24. Jian Lu, Yansun Xu, J. B. Weaver and D. M. Healy, Jr., *Noise reduction by constrained reconstructions in the wavelet-transform domain*. Department of Mathematics, Dart-

mouth University, 1992.

25. S. Mallat and W. L. Hwang, "Singularity detection and processing with wavelets", *IEEE Trans. Info. Theory.* **38**, 2 (1992) 617-643.

26. Y. Meyer, *Ondelettes et opérateurs I: Ondelettes*, Hermann, Paris, 1990.

27. P. Moulin, "Wavelets as a regularization technique for spectral density estimation", in *Time-Frequency and Time-Scale Analysis*, IEEE, New York, 1992, 73-76.

28. E. P. Simoncelli, W. T. Freeman, E. H. Adelson and D. J. Heeger, "Shiftable multiscale transforms", *IEEE Trans. Info. Theory* **38**, 2, 587-607.

29. C. Stein, "Estimation of the mean of a multivariate normal distribution", *Ann. Statist.* **9** (1981) 1135-1151.

30. G. Wahba and S. Wold, "A completely automatic French curve", *Commun. Statist.* **4** (1975) 1-17.

31. G. Wahba, "Automatic smoothing of the log periodogram", *J. Amer. Statist. Assn.* **75** (1980) 122-132.

DEPARTMENT OF STATISTICS, STANFORD UNIVERSITY, STANFORD, CA

E-mail: donoho@playfair.stanford.edu

Recent Titles in This Series

(*Continued from the front of this publication*)

19 **J. T. Schwartz, editor,** Mathematical aspects of computer science (New York City, April 1966)

18 **H. Grad, editor,** Magneto-fluid and plasma dynamics (New York City, April 1965)

17 **R. Finn, editor,** Applications of nonlinear partial differential equations in mathematical physics (New York City, April 1964)

16 **R. Bellman, editor,** Stochastic processes in mathematical physics and engineering (New York City, April 1963)

15 **N. C. Metropolis, A. H. Taub, J. Todd, and C. B. Tompkins, editors,** Experimental arithmetic, high speed computing, and mathematics (Atlantic City and Chicago, April 1962)

14 **R. Bellman, editor,** Mathematical problems in the biological sciences (New York City, April 1961)

13 **R. Bellman, G. Birkhoff, and C. C. Lin, editors,** Hydrodynamic instability (New York City, April 1960)

12 **R. Jakobson, editor,** Structure of language and its mathematical aspects (New York City, April 1960)

11 **G. Birkhoff and E. P. Wigner, editors,** Nuclear reactor theory (New York City, April 1959)

10 **R. Bellman and M. Hall, Jr., editors,** Combinatorial analysis (New York University, April 1957)

9 **G. Birkhoff and R. E. Langer, editors,** Orbit theory (Columbia University, April 1958)

8 **L. M. Graves, editor,** Calculus of variations and its applications (University of Chicago, April 1956)

7 **L. A. MacColl, editor,** Applied probability (Polytechnic Institute of Brooklyn, April 1955)

6 **J. H. Curtiss, editor,** Numerical analysis (Santa Monica City College, August 1953)

5 **A. E. Heins, editor,** Wave motion and vibration theory (Carnegie Institute of Technology, June 1952)

4 **M. H. Martin, editor,** Fluid dynamics (University of Maryland, June 1951)

3 **R. V. Churchill, editor,** Elasticity (University of Michigan, June 1949)

2 **A. H. Taub, editor,** Electromagnetic theory (Massachusetts Institute of Technology, July 1948)

1 **E. Reissner, editor,** Non-linear problems in mechanics of continua (Brown University, August 1947)

ISBN 0-8218-5503-4

9 780821 855034